LIME WORKS

Patrick McAfee

commissioned by

The Building Limes Forum Ireland

Æ

ASSOCIATED EDITIONS

Dedicated to the memory of
James O'Callaghan and Dick Oram

Published 2009 by
The Building Limes Forum of Ireland
www.buildinglimesforumireland.com

and

Associated Editions
33 Melrose Avenue, Dublin 3
www.associatededitions.ie

Distributed by Associated Editions

ISBN 978-1-906429-08-9

Editorial Consultant: Roberta Reeners
Design and Production: Vermillion Design
Index: Hugh Brazier
Print: Nicholson & Bass

The practice of using lime and its associated technology are constantly
changing and developing. Every building or structure differs in its
construction, problems and requirements. While every endeavour was made
to ensure the information in this text was correct at the time of writing, those
working with lime are advised to keep abreast of changes and to consult with
conservation and other relevant building experts for professional advice on
the use of lime for their specific needs and circumstances.

Front cover:
Kildrought House,
Celbridge, County Kildare.
Back cover: Farm House,
Carrickspringan, County Meath.
Boreen Bradach, Kinnegad,
County Westmeath.

CONTENTS

FOREWORD

Lime Works emerged from the desire of conservation and environment organisations to provide information on the use of lime both in historic and new buildings in Ireland. A traditional building material used throughout the island for many hundreds of years, lime was almost abandoned during the latter part of the twentieth century in favour of mortars and plasters made from cement. As the practical and aesthetic advantages of using lime became more widely recognised, recent years have witnessed a revival in its appreciation and use.

Published by the Building Limes Forum of Ireland with the support of many organisations and individuals, *Lime Works* provides easy-to-read, easy to understand and easy-to-use information on the many uses of lime in building, including the repair of historic structures and in new buildings. The author, Patrick McAfee, is unrivalled in the depth of his experience and expertise in working with, teaching and writing about the use of lime in Ireland.

Using a question-and-answer format, *Lime Works* explains why lime is important when repairing historic buildings. It also addresses the environmental benefits of using lime – an issue which has seen a growing interest in the use of lime as a sustainable building material. *Lime Works* is lavishly illustrated with many contemporary images, which impart the information in a clear and straightforward way. Aimed at the building owner, the practitioner and the specifier, the book is divided into three separate sections, with numerous cross-references which offer the reader easy access to all information.

We hope that *Lime Works* will benefit our built heritage in the information it provides on best conservation practice. With buildings illustrated from all parts of the island of Ireland, it will also raise awareness of the enduring relevance of this most traditional of building materials and help to advance the use of lime in sustainable building.

For further information on the Building Limes Forum Ireland, please consult www.buildinglimesforumireland.com

Ivor Mc Elveen
Chairman
Building Limes Forum Ireland

ACKNOWLEDGEMENTS

This publication would not have been possible without the generous support and financial assistance from a very large number of people and organisations.

In particular, the author, Patrick McAfee, and the Building Limes Forum Ireland would like to thank the following:

The BLFI Publication and Editorial Committee
Gráinne Shaffrey (Chair), Mary Hanna, James Howley, Dan McPolin and Grellan D. Rourke.

Those who provided financial support
The Environment Fund – Department of the Environment, Heritage and Local Government

The Heritage Council; Northern Ireland Environment Agency; Donegal County Council: part-funded under the County Donegal Heritage Plan (2007–2011); Fingal County Council; Kildare County Council; Kilkenny County Council; Association of Architectural Conservation Officers (AACO).
Edward and Primrose Wilson, Michael Collins Architects, Nolans MPC.

Those who contributed to the content, read drafts and provided invaluable feedback and support
The main committee and individual members of BLFI, both past and present, Liam Duffy, Ian Brocklebank, Kevin Holbrook, Dr Gerard Lynch, Jeff Orton, Richard Smith and the BLF UK Committee.

The Editorial and Design Team
Roberta Reeners, editor, and Anne Brady and the team at Vermillion Design.

Those who contributed to the photographic content
Con Brogan, Cork County Council, Edward Byrne, Kevin Carrigan, Dermot Collier, John Cotter, David Davidson, Denis Deasy, Ana Dolan Hugh Dorrian, James Fraher, Richard Good-Stephenson, Mary Hanna, Sean Harrington, Kevin Holbrook, James Howley, Irish Landmark Trust, Chuck Leese, Limerick County Council, Brian McAfee, Patrick McAfee, Alan McGrath, John McNamara, Donal Murphy, Dan McPolin, Office of Public Works, Peter Pearson, Grellan D. Rourke, Gráinne Shaffrey, Henry Snell, Henry Thompson, Pierce Tynan, David and Sinead Tyner, Amanda and Jack Louis Wilton.

Thanks also to Maurice Brian, Terry Fitzpatrick and Michael Riordan of FÁS, and Gerard McAfee.

To those who have contributed to the revival of lime use in Ireland over the last number of decades and who, in doing so, have added to our greater appreciation and understanding of its nature and application.

Comhshaol, Oidhreacht agus Rialtas Áitiúil
Environment, Heritage and Local Government

Comhairle Contae Fhine Gall
Fingal County Council

Opposite: Wet dash render, Dovecote, Dublin

PART 1

THE BUILDING
OWNER

INTRODUCTION

As a building owner, you may know that repairs to old buildings should be carried out using lime. As your building was originally built, plastered and rendered in lime, lime should be used when repairs are necessary. With insufficient knowledge of the subject, you decide to explore it further. Information is not plentiful and in many cases can be downright confusing. You read, email, and ask questions, and begin to gain knowledge, to become a lime enthusiast.

Old buildings need such enthusiasts. In the last twenty years, they have assisted greatly in bringing about change by questioning and demanding that lime be used by those involved in the repair of old buildings. Not so long ago, old buildings were repaired with cement-based mortars and concrete, which has created many problems. It is only in recent years that knowledge about lime mortars has been rediscovered, and lime products and skills have become more widely available again. With continual research and practice, our knowledge of lime is growing and changing. Who would

Limewash over stone. Clare Island, County Mayo

Interior of Salterbridge Gate Lodge after restoration

Top left: *Before*: a gate lodge in ruins. Salterbridge Gate Lodge, Cappoquin, County Waterford

Top right: *After*: the gate lodge restored using lime

have thought that it should take so long to rediscover what was once common practice in the past, the knowledge of which is so useful when we attempt to repair old buildings today? Lime is now even being used without cement to construct new buildings as it was for thousands of years before the arrival of cement in the mid nineteenth century.

The Building Limes Forum Ireland (BLFI) provides a forum for everyone who is interested in lime, giving them a chance to come together and discuss, demonstrate and research the use of lime for many different purposes.

Commercial companies, government bodies, churches, a client in the process of having a building repaired – all want the best for the old building in their care, and lime is almost always the correct answer.

As the same questions about lime are asked by many building owners again and again, it has proved helpful to take the most commonly asked questions and try to answer them, illustrating the principles by using drawings and photos to make the whole thing more understandable. In this way, your voyage of discovery may be shortened.

Some questions and answers may overlap and repeat recurrent issues, but this will help you in the process of learning more about this fascinating subject.

More detailed and advanced aspects of lime are covered in the Practitioners and Specifiers sections of this book.

Now, let the journey begin.

WHAT IS LIME?

Lime is derived from limestone, a sedimentary rock formed from the shells or skeletons of prehistoric creatures. Limestone, chalk and shells are composed of calcium carbonate $(CaCO_3)$. When burned in a kiln at high temperatures, they produce lime.

The Irish language refers to limestone as *cloch aoill* or manure stone. This indicates that its primary use was in agriculture, as it was applied to the land to increase the alkalinity of acid soils. Its secondary use was in building. In Ireland, the use of lime for building lasted over a thousand years until Portland cement began to dominate the market, particularly from the mid twentieth century onwards. Lime is now making a comeback in both the repair of old buildings and in new-build.

Top: On being burnt in a kiln, limestone (left) produces quicklime (centre). Quicklime added to water produces lime putty (top right), while a smaller amount of water will produce a hydrate (bottom right)

Top right: Painting of an old Irish lime kiln

Right: A typical nineteenth-century lime kiln. Ballindoolin, County Kildare

Lime mortar used to build a
traditional stone wall

Natural hydraulic lime mortar being
used to build modern brickwork.
Balgaddy, County Dublin

LIME USE

Q. How is lime used in building?

A. Lime is used as a *binder* in the following instances:
- Mortar for building stone or brick masonry
- Mortar for pointing joints between bricks, stones and earth
- Plaster for internal plastering of flat surfaces, decorative mouldings and features
- Render for external wall surface finishes
- Limecrete (lime and coarse aggregate) for ground floors
- Limewash
- In combination with bio aggregates such as hemp
- Earth structures

LIME MORTAR COMPOSITION

Q. What is lime mortar composed of?

A. Lime mortar is composed of three main ingredients:
- Lime
- Aggregate
- Water

While these are the basic ingredients, it is important to remember that there are different types of lime. In addition, other ingredients can be added to lime to enhance its ability to set in water, reduce setting times, or increase compressive strength.

WHY USE LIME?

Q. Why should I use lime to repair my old building?

A. Here are some of the most important reasons for using lime.
- 'Like-with-like' is an expression often used when it comes to repair. In this case, a structure built with lime is best repaired with lime. Mortars made with lime have characteristics which are different from those made with cement.

Lime, sand and water

- Lime mortars are used for building and pointing, external renders, internal plasters and limewashes. They are permeable – they allow solid masonry structures to breathe, so water that gets in can get back out again through evaporation.

- Movement takes place within flexible lime mortar joints, but not in the more rigid masonry units of stone and brick. This movement reduces serious cracking in these more rigid materials. Modern cement-built structures are rigid and often have to incorporate unsightly expansion joints to accommodate movement.

- There are ecological benefits in using lime instead of cement. As lower temperatures are required to produce lime in the kiln, less energy is expended in its manufacture. The carbon dioxide produced in burning limestone is re-absorbed as the lime hardens or sets through a process known as *carbonation*. With cement, the carbon dioxide produced in its manufacture is not re-absorbed by the cement.

- Because lime mortars do not set as hard as cement mortars, brick and stone are recyclable from lime-built structures. Recycling building materials for re-use is an important factor in reducing waste – and one of the main reasons behind the growing use of lime in new masonry structures.

- Interventions to old buildings using lime can generally be reversed without damaging the original fabric, an important principle in conservation.

- Lime mortars contain *free lime* which can close or *self heal*. This is called *autogenous healing*. Water transports free lime or un-carbonated lime where it can carbonate and heal or close narrow cracks.

- Lime mortars and plasters are *hygroscopic*. They absorb moisture from the atmosphere of the internal environment of the building, thereby reducing condensation and providing a healthier living environment. This is also good for the general fabric of the building.

- Using lime creates less waste. The shelf-life of all limes is longer than cement. *High calcium lime* (see 'Lime – Types', page 10) can be stored indefinitely as a lime putty. Lime mortars set more slowly and so are useable many hours, or sometimes even days, after being mixed.

- Lime mortars are more workable than cement mortars. For the stonemason, bricklayer or plasterer, this is important in achieving both quality and efficiency.

- Aesthetically, lime mortars enhance the varying natural colours in brick and stone. Lime renders and plasters have softer textures, and these are further enhanced when a *limewash* is used as a protective, decorative finish.

- Lime in combination with hemp as a bio aggregate has a negative carbon footprint. When growing, hemp absorbs carbon dioxide through photosynthesis. The carbon dioxide is sequestered when the hemp is used as an aggregate with lime. Hemp in combination with lime provides a sustainable, low embodied energy, high insulation and healthy living environment. Lime is also being used with straw, coppiced timber and recycled glass.

Ochre-coloured limewash, wet dash render, rubble stone and dimension limestone reveals at openings. Ballindoolin, County Kildare

LIME – TYPES

Q. It seems that there are different types of lime, with confusing names. Can you explain these different types?

A. There are different types of lime, and some limes have several different names. The following summary will help. The two types of lime that you are likely to encounter are:
- High calcium lime
- Natural hydraulic lime (NHL)

High calcium lime is also called non-hydraulic lime, fat lime, air lime and CL90. The term *high calcium lime* is used throughout this book.

Pure or nearly pure calcium carbonate (limestone) produces high calcium lime, a lime that requires access to the carbon dioxide in air in order to harden.

High calcium lime is generally used as a lime putty (calcium hydroxide) for the conservation and repair of old buildings. Lime putty is commonly used with sand for internal plastering on old buildings.

Lime putty has the consistency of soft cheese and is usually sold in plastic tubs. When used, it hardens slowly by absorbing carbon dioxide from the air. It will not harden in continuously damp situations or underground where it does not have access to carbon dioxide.

For over 2,000 years, *pozzolans* (volcanic ash or brick dust) have been added to high calcium lime to increase its *hydraulicity* (the ability to set in water), accelerate its setting time, and increase its compressive strength, if required.

High calcium lime is also available as a powder referred to as a *hydrate* (in this case, calcium hydrate) and is sold in paper sacks. Calcium hydrate is normally used only in modern construction to increase the workability of cement mortars.

The second basic type of lime is natural hydraulic lime (NHL). It is produced by burning argillaceous limestone (limestone containing clay) and/or siliceous limestone (limestone containing silica). On being burnt, these limestones produce reactive aluminates and silicates that

Mature lime putty on top of a coarse sand

High calcium lime

Natural hydraulic lime. NHL1 is only
recently available

cause a hydraulic set when in contact with calcium hydroxide. Natural hydraulic lime comes in four grades and is able to set in wet or damp conditions. It also sets faster and reaches a higher compressive strength than high calcium lime.

If natural hydraulic lime is used where it has access to air, carbonation will occur in addition to the hydraulic set. This is more pronounced in the weaker grades of NHL.

Natural hydraulic limes are commercially available only in a powdered or hydrated form, delivered to the site in a paper sack or mortar silo.

SAND CHOICE

Q. How important is the choice of sand in the partial rebuild of my rubble stone barn?

A. Rubble stone consists of stones which are irregular in shape and size. It is usually laid as-found, or roughly shaped with a hammer. Joint sizes often vary, with small pinning stones inserted between the larger ones. Rubble stone walls can vary greatly in quality.

Limestone rubble laid with a lime
mortar and a coarse sand

A coarse sand

A close look at the original mortars in such work usually reveals that a coarse sand was used, sometimes bordering on a light gravel. Through weathering, sand particles become exposed, not only on mortars but also on external renders. Aesthetically, this adds wonderful colour and texture.

In general, sands which are commonly available for modern brickwork, blockwork and plastering are too fine for use in the repair of traditionally built rubble stone structures. It may take some time, but it is always worth sourcing a local sand that approximates the original in colour, shape and size.

MISGUIDED 'IMPROVEMENTS'

Q. We live on an island off the west coast of Ireland, and our house has developed damp spots on the inside face of the outer walls. My mother has lived here all her life and has never seen anything like it. The house is also noticeably colder in winter. Four years ago, we spent a considerable amount of money on improvements. The external lime render and internal lime plaster were removed, and the walls were pointed with sand and cement. While we have a most attractive stone house, we now have damp problems. The recent application of a clear, waterproofing liquid to the external stone walls has not helped. If anything, it seems to have made the damp worse.

A. You have committed nearly all the errors one can make with an old stone building, in the mistaken belief that your 'improvements' would enhance the weathering performance of the building envelope. It will be difficult to undo what has been done.

In the past, houses were rendered with lime and sand to protect them from the elements. However, these lime renders were not waterproof. In Ireland, we usually experience intermittent rain patterns, with the wind blowing between showers. So while the rain gets in, lime mortars and renders allow it to evaporate outwards before it reaches the inside face of external walls.

With continuous rain over a prolonged period, damp patches may appear internally in lime-coated buildings but this is rare. In some parts of Ireland, the rain comes in horizontally off the sea, with walls facing south to north-west getting most of the impact. Lime renders and mortars assist the evaporation of any rain that gets into these structures. Lime mortar joints between stones assist much of this evaporation, in effect acting like drains. Remember, your walls are solid, not the modern cavity type. When rain gets in, it must evaporate to the outside.

Stripping renders, pointing with impervious sand and cement, and applying a waterproof liquid – as you have done – will cause problems. Water will still find its way in but will be prevented from getting out again.

What is the solution? Should the sand and cement pointing be removed? Can it be removed, or is it near impossible to do so because it is so hard? In removing it, will the stone and brick be damaged? Can the waterproofing liquid be removed? If so, how can this be done?

As a general rule, if sand and cement pointing or render are creating serious problems, they are best removed and replaced with lime. Water repellents are generally short lived and rarely effective.

Deciding what to do now is best left to a conservation architect or engineer specialising in the repair of stone structures.

MODERNISING AN OLD BUILDING

Q. I have a lot of work ahead of me. As an old building can be more expensive to modernise than buying a new one, the house I am thinking of buying would probably be best demolished. It will be expensive to straighten walls, replace sliding sash windows, and dry-line the internal walls with plasterboard. Uneven floors, ceilings and a twisted staircase all need replacing. This means that there will only be a basic shell of four walls by the time I'm finished. Have you any thoughts on how to modernise such old houses, and should I use lime to do so?

A. For old houses to be made habitable, they must provide a reasonable level of comfort. Renovations should, however, be carried out with respect for the building.

Above: A building such as this eighteenth-century farmhouse must be repaired with respect for what is there – the materials, setting and style. Mayglass, County Wexford, prior to repair

Right: Sensitive repairs have been carried out at Mayglass, County Wexford

To over-modernise such structures is to assist in their disappearance from the landscape. Straightening walls, replacing windows, levelling floors and dry-lining the interior would result in a loss of character – the building should be allowed to retain the characteristics of ageing.

It is possible to add sensitively to an existing building in a subtle manner or in a contemporary style. You must always endeavour to maintain the integrity and personality of the old building, however. You should also try to work with the prevailing landscape, and think about the relationship between the existing building, any new addition and the surrounding spaces.

A worn threshold, a sloping floor, original timber windows – all honest signs of ageing – are often the very features that attract us to old homes in the first place. We do not need to achieve 'new-build' standards of finish in the renovation of old houses.

Old buildings have embedded energy – for example in the fuel used to produce the lime, the human energy expended to quarry, transport and lay stone. Preserving and reusing old buildings maintain this energy resource, reducing the overall energy output necessary to replace and rebuild.

Old buildings do not suit everyone, of course. What they need are owners who are in sympathy with their nature. Old houses suit some, while more modern houses suit others. It may be best to consider if owning and repairing such an old house is best for your situation.

PERIODIC SEVERE WEATHER

Q. After a recent very heavy rain, leaks and damp patches were discovered on the inside walls of our local stone church. These were particularly obvious at window openings and on the undersides of arches. While similar damage occurred about twenty years ago in a similar downpour, the church has otherwise been dry. We are now eager to get started on lime pointing work. What is the next step?

A. You may have to do nothing at all. Periodic severe weather conditions can cause these problems, usually at window and door *reveals* (the side walls of window and door openings which visibly show the thickness of the wall) and under arches and lintels. Exceptionally heavy and prolonged rain combined with wind will drive rain into the masonry structure. Generally, this will first be noticed where there are openings in the walls.

It is unlikely that pointing will solve these problems. The fact that this last happened twenty years ago bodes well for the future. Any removal and replacement of mortar between stones is time-consuming and expensive. And instead of making matters better, it could make them worse. Unless there are obvious open mortar joints and holes, with vegetation growing on the wall surface and in mortar joints, it is best to leave well enough alone. Consult a conservation specialist for further advice.

MILL HOUSE

Q. I have been told that all the repairs to my old four-storey stone mill building must be carried out with lime. At some time in the past, a large hole was knocked through one of the walls so that large mill stones and machinery could be removed. There are some smaller holes where stones have fallen out of the walls as a result of ivy growth. There is also severe mortar loss between stones on the south and west faces. Here and there on the east and north elevations, there are the remains of a rough lime render. I would like to know how to proceed with the repairs.

A. As this work requires professional analysis, you should consult an architect or engineer who specialises in conservation. He or she will take you through all the necessary stages and requirements relating to this type of work, including how to find a suitably experienced contractor to carry out the work.

The existing mortars should be examined, and replacement mortars specified. A range of lime mortars may be required for different areas of the work. It is possible that, in addition to pointing, the mill may have to be *wet dashed* (a *thrown-on* external lime render sometimes called *harling*) using lime. Decisions like this should only be made by a specialist.

POINTING – INDICATORS

Q. How can I tell if my traditional rubble stone house with its original mortar needs to be pointed?

A. The following basic signs will indicate whether your house requires pointing or not (refer to pages 66 and 185 for more detailed information).

- Noticeable mortar loss from joints between stones
- Loss of pinning (small) stones from larger mortar joints

The loss of mortar and stone pinnings is more common on wall surfaces which face prevailing wind and rain. Parapets are particularly vulnerable. Selective pointing is best, leaving areas with sound mortar alone. Pinning stones should be replaced. A lime mortar should be used which matches the original.

LIME – INSULATION

Q. I believe that a lime plaster will provide sufficient insulation to keep my solid rubble stone house warm. I bought the house two years ago and find it very cold during the winter. The external walls are bare stone. There is no external render or internal plaster.

A granite rubble wall recently pointed with lime mortar. Its pinnings (small stones) have all been reinstated. Small Cone, Killiney Hill, County Dublin

A. Your house was probably rendered and plastered originally. These would have provided a certain degree of insulation, but not sufficient to meet modern standards.

Owners sometimes strip their walls of external lime render and internal plaster to expose the stone beneath, intending to make the house more appealing. In doing so, they create serious problems for themselves and future owners.

Plain lime plasters and renders may provide only minor insulation to solid stone walls. However, they also reduce condensation, are aesthetically pleasing, historically in keeping, and protect lime mortar or mud mortar joints between stones from getting washed out.

In the past, fires were kept burning in houses for both cooking and heating. Although the walls were not insulated, they were built of solid, thick stone that acted to some extent as heat banks, holding and returning heat at a low level within the internal environment of the building. Windows were small, and roofs were often thatched, both of which reduced heat loss. Modern living generally requires a building to be heated quickly at the turn of a switch, something that is difficult to achieve in older buildings.

It may be worth considering using a lime and hemp aggregate mix for plastering the inside face of the external walls. Hemp can also be used with lime in floors and as a lightweight insulation in lofts. Specialist advice should be sought on the use of this product.

Internally, wainscoting the bottom half of a wall using timber boarding was once common. This is another possibility, and high-grade modern insulation could be concealed behind the wainscoting. Your architect will advise on what insulation to use. It will be necessary to ventilate the space behind the wainscoting.

Externally, a render which replicates the original would also be advantageous if you can find out what that was. An external render with good evaporation characteristics will be required. This will reduce the risk of interstitial

A derelict nineteenth-century barn under restoration at Cronykeery, Ashford, County Wicklow. The walls, which are a mixture of granite and slate, have been pointed using natural hydraulic lime and a new roof installed with a solar heating panel

Interior of the same barn with a new limecrete floor

condensation within the solid stone wall where insulation has been used internally.

A constant low level of heat, particularly during the winter months, will work well in traditional solid-wall buildings.

Above: A brick radiator under construction, using brick and lime mortar. The lime mortar allows thermal movement, preventing cracking. Johnston Central Library, County Cavan

Above Right: Brick radiator upon completion

DRY STONE WALL COTTAGE

Q. We are in the process of having our seaside cottage pointed, and it is taking an inordinate amount of time. We removed the old lime render and are using the recommended lime mortar. What we don't understand is that there does not seem to be an original mortar of any kind between the stones. We find that we are using large quantities of mortar to complete the work.

A. You probably have a *dry stone* structure that was later rendered with lime to protect it from the weather. When it was constructed, lime may not have been available in that area, or it could have been beyond the means of the original owner. Nobody knows how many dry stone houses survive today, as the original stone may now be hidden by the subsequent application of lime renders and plasters. In the past, lime or mud were commonly used as mortars, but dry stone houses were built too.

A vernacular cottage in ruins,
built in dry stone and later pointed,
thinly rendered and limewashed
externally. Aran Islands,
County Galway

Now that you have started, it is probably best to
continue – but you should certainly consider reinstating
the lime render. Many void spaces will exist within the wall
structure that may cause problems later. This is because
pointing without lime mortar grouting (a specialist
process) will not do the job adequately. Pointing should be
inserted as deeply as possible and compressed with a
ramming iron.

It is important that the lime being used for pointing
and rendering is sufficiently permeable so that any water
that gets into the structure can get back out again.

POINTING ASHLAR STONE

Q. I am responsible for the care of our local nineteenth-century church. The stone in the church is cut finely with very small joints. Some of these joints look dry. It appears that the mortar has been washed out by rain. It has been recommended that these joints be widened by cutting, as it is impossible to get mortar into them. I have no idea what mortar should be used. Before we commence this expensive work, I am looking for advice.

A. Your church was constructed with ashlar stone, a term that denotes well-cut *dimension stone*, of any type, with small mortar joints. The joints can be 3mm or less. The accurate cutting and surface dressing were all carried out by a skilled stonecutter who specialised in cutting but not the laying of stone. The stones were laid by skilled stonemasons using lime mortar. Stonemasons built stone structures and carried out basic cutting.

Because the stones were laid with such fine joints, the sand in the mortar had to be very fine. We can sometimes see fine sand in the lime mortar in joints such as these, as well as evidence that lime was used without sand. The trouble with using lime without sand is that it is likely to shrink, turn to powder, and get washed out.

Ashlar limestone displaying the art of the stonecutter and the stonemason, recently repointed using lime. Christ Church Cathedral, Dublin

Getting mortar into ashlar joints is not easy – but it can be done. Widening these joints with an angle grinder or a con saw should not be considered under any circumstances. This would destroy historical aspects of the building, resulting in serious, irreversible damage to its appearance and performance.

This work requires professional advice to start with. Then if it is decided to go ahead with the work, a specialist contractor will be required. Otherwise, do not attempt it.

LIME – WATER-PROOFING

Q. I've been told to point the external stone walls of my house with a lime mortar mix. Will this waterproof my house?

A. No. Water will inevitably get through the external wall face, and it is essential that it gets back out through this face. Otherwise, dampness will show on the internal walls. Much evaporation takes place through mortar joints. Most lime mortars are permeable and assist this process.

LIME PLASTER CEILINGS

Q. Because they attract dry rot and vermin, I have been told to remove my old lath and plaster ceilings and replace them with modern plaster board. Personally, I like their appearance. Although they are two hundred years old and slightly uneven, they seem sound to me. If I removed them, I would also lose the lovely cornice between the ceiling and the wall.

A. Follow your instinct – you are right. There is a general belief that ceilings like this should be removed – but nothing could be further from the truth. If needed, repairs are certainly possible. Holes can be filled, and dropped sections screwed or tied back. Originally, the cornice would have been run *in situ* (created in place). The plaster was composed of lime and sand, with animal hair added as reinforcement. Its presence reflects the available material and skills of the craftspeople who created it, the social standing of the client, and the building style of the time. It is therefore special and should be retained.

Mud plasterwork. Kilmallock,
County Limerick

MUD PLASTER

Q. The first coat of plaster on the inside walls of my basement is yellow in colour and looks like earth. The two coats of plaster over this are lime and sand. The question is: what is the first coat?

A. The first coat on your basement walls may be mud. Because the walls were in contact with earth on their exterior face, the original intent was to waterproof them on the internal face. This is called *dóib bhuí* (yellow daub or mud) which was well known as a building material for many centuries. The muds were not always yellow, and were sometimes blue/grey or brown. They were resistant to the ingress of water. If the mud plaster on your basement walls is serving its original purpose, leave it alone.

A semi-ruinous farmhouse in need of repair. Carrickspringan, County Meath

The same farmhouse after repair using lime

LIMEWASH

Q. The old outbuildings and barns on my property are all limewashed white. Why was this so popular in the past?

A. Limewash was used as a disinfectant in cattle byres and is still commonly used today for that purpose. It was also strongly associated with human habitation for similar purposes in an attempt to create healthier living conditions. It also looks attractive and brightens up a property.

Many vernacular buildings had small windows, so internal limewashing was useful to reflect internal light.

The application of a limewash had other benefits too. It reintroduced lime-rich water into renders which then carbonated. This preserved and extended the life of these renders.

Cottages limewashed in Yellow Ochre and plain white. Kilmallock, County Limerick

The remaining traces of old coloured limewashes provide valuable information for future repairs

LIMEWASH COLOUR

Q. One of the coats of limewash on my house is pink-to-red in colour. Another is yellow. What colour types are these?

A. The pink/red colour is probably *Venetian Red*. It is an iron oxide that was commonly added to limewashes to create a vibrant red colour. With weathering, this diminishes over time to pink. Venetian Red pigment is still readily available today.

The yellow colour is possibly *Yellow Ochre, Raw Sienna* or *Ferrous Sulphate*. Applied properly to suitable backgrounds in a sheltered environment, limewashes can have a long life.

White limewash was sometimes enhanced further by the addition of a *ball of blue* (blue bag), the same substance used in laundries to create a startling white finish.

LIMEWASHES AND DISTEMPERS

Q. I intend to repair my late nineteenth-century thatched vernacular stone cottage using lime as an external render (wet dash) and as an internal plaster. What finishes can I apply to these surfaces to maintain the benefits of using lime that would also be in keeping with this style of architecture?

A. At the time your cottage was built, the decorative surface finishes were few. The external render was commonly limewashed. A simple limewash is composed of lime putty and water, with or without a colour. After multiple coats, this rounded off the aggregate in the wet dash, creating interesting textures that are pleasing to the eye. While it was often carried out once a year, when carried out properly and in sheltered situations, every five years was possible. The limewashing of external lime renders is often seen as an essential part of the overall process of weathering by increasing the life span of the render. You should therefore strongly consider limewashing your external render.

The internal plaster was either left undecorated or finished with limewash or a soft *distemper* paint. Distemper is composed of chalk (whiting), an earth pigment, and size (animal glue) or casein, and is available in tins. Limewashing is also a successful internal finish for a vernacular cottage. It is worth noting that limewash should not be applied over an existing impervious surface finish such as paint, sand and cement etc. because it needs to absorb into the background surface. Limewash will not mask damp patches on internal wall surfaces. Fortunately, it allows evaporation to take place rapidly. Limewashed or distempered surfaces cannot be cleaned by washing.

WEEPING STONES

Q. I have been told that I have weeping stones on the internal face of the external walls of my house, and that these will allow rainwater to come through the wall. I notice damp spots every so often on the exposed internal wall faces. What are weeping stones? Should I remove them, as advised? Will lime prevent this from recurring?

A. A weeping stone on the inside face of an external wall is usually the result of condensation on the surface of that stone rather than rain passing through the thickness of the wall. Any hard, dense, smooth, exposed, impermeable stone is liable to show this phenomenon. *Through stones* (stones used to tie both faces of a wall together and therefore passing through the full thickness of the wall) are more liable to suffer condensation because they are acting as *cold bridges*, transferring the outside temperature to the internal wall face.

Applying a lime plaster or even a limewash will improve the situation because lime is *hygroscopic* (absorbs and slowly releases moisture). This is one very good reason not to remove original plasters to expose stone.

The removal of weeping stones, particularly if they are through stones, is neither practical nor sensible.

BRICK VAULT – DUST

Q. We have a restaurant with an early twentieth-century brick vault that is quite attractive, although built with a rather rough red brick and large mortar joints. The trouble is that dust continually falls from the vault. This looks like a lime mortar. There are also tiny particles of brick. For a considerable time, there was a leaking roof above the vault. We have been advised to seal the vault surface underneath with a chemical sealer. Is there any alternative?

A. Time may alleviate the problem and it may be best to do nothing at all. If the roof has been leaking, the vault may have become very damp, taking in a considerable amount of water. This dampness is now drying out, and the dust you see is a consequence of this. Once the vault is dry and there are no more wet/dry cycles, the problem should be alleviated. An occasional light brushing and/or vacuuming may be all that is required.

If the problem persists, you could consider using a limewash (lime putty and water). The limewash should consolidate most of the brick and mortar joint surfaces, thereby reducing the problem further. (A limewash will cover but not seal the surface and will allow water to continue to evaporate.) This will work more successfully if the bricks are porous rather than machine-pressed with a dense, smooth surface. However, the limewash will also hide the natural and variable colour of the bricks themselves.

Apply the limewash in multiple thin coats. You will probably experience white salts appearing through the limewash at isolated places. This indicates that the vaults are still drying out. Light brushing and an occasional further coat of limewash may be required.

BRICK CHIMNEY REBUILD

Q. My brick chimney seems to have expanded both horizontally and vertically over the years. It is now in a dangerous state and liable to collapse. The old red and orange clay bricks are disintegrating. I don't want to create the same problem again when I rebuild it. How should I go about this, and what mortar should I use?

A. Expansion and contraction from heat and cold, sulphate attack from soot and water, and the fact that a chimney is exposed to very severe conditions caused by wind, rain, frost and sun, have led to these problems, which are common in old stacks.

Bricks used for chimney construction should be hard and well burnt. With older red bricks, it is the darker-coloured bricks that are the most durable. The lighter red and orange bricks were often under-fired and too soft. You need a skilled bricklayer to rebuild your chimney.

A nineteenth-century chimney, showing deterioration from previous inappropriate repair using cement. Kilmallock, County Limerick

A lime mortar that will match the chosen brick type is best, taking into account the severe weather location of the chimney and future sulphate conditions. The new chimney should reflect the size, shape and style of the old chimney. Due regard should be given to choosing the replacement bricks.

Clay flue liner pots will reduce future sulphate attack. If a flue liner cannot be inserted, the inside of the flue should be parged (plastered) with a lime mortar.

SULPHATE STAINING

Q. There are ugly brown stains on the fire breast above where the fireplace used to be. What is this, and is there anything that can be done about it?

A. This stain is caused by sulphates. In this case, sulphates are created when water and soot get together. It is very corrosive, attacking iron, soft brick, some stones, lime mortar and plaster. Stains vary from yellow to orange/brown/tobacco in colour.

Sulphates created by soot and water can damage brick on fire breasts and chimneys, creating a fire hazard. The inside of the flue was originally parged with lime, sand and cow manure. This reduced sulphates and the risk of fire, but has been lost over time. Ballitore, County Kildare

Originally, the internal flue was parged with lime, sand and cow manure. The cow manure has properties which resist sulphate action. This parging has probably long since gone and is now impossible to replace, as it would require access to the flue.

Your reference to 'where the fireplace used to be' indicates that it has been blocked up. A vent should be installed in the blocked-up fireplace, as low as possible. This will ensure a draft of air through the chimney, alleviating dampness and sulphate attack. Covering the chimney top, with allowance for ventilation, may also be worth considering.

Then, when the fire breast is dry, it can be replastered. Sulphates in the fire breast will continue to emerge on the wall surface over time. If fresh cow dung is added to the lime and sand plaster mix when the stained area of the wall is replastered, there is a good chance the staining will not recur for some time (see 'Bricklaying', page 114).

NEW-BUILD USING LIME

Q. As owners of a large vacant site on the outskirts of a city, we intend to build three-storey apartments using brick, with some stone on the external elevations. Concrete block will be used internally. Rather than using a sand and cement mortar, is it possible to use lime? If so, what are the advantages?

In the first phase of this social housing scheme, 55 houses were built using concrete blocks laid in lime mortar. The finish was natural hydraulic lime and sand. Both roughcast and trowelled finishes were used. Boreen Bradach, Kinnegad, County Westmeath

A. In recent times, natural hydraulic lime has made a comeback for new-build. The advantages of using the correct lime mortar are:

- Possible elimination of unsightly expansion/ movement joints.
- Lower energy input in the manufacture of lime compared to cement.
- Materials like brick, stone and concrete block are more recyclable when built in lime mortar.
- Lime mortars often look better than cement mortars.

Lime in new-build can be used in mortars, renders, plasters and limecrete on ground floors.

A large modern public building built using lime. Johnston Central Library and Farnham Centre, Cavan

Brick openings. Johnston Central Library and Farnham Centre, Cavan

POINTING BRICK

Q. Our public authority has offices in an eighteenth-century Georgian brick building in Dublin. Any remaining pointing consists of a very thin projecting bead of white mortar, with red mortar each side flush with the brick. The building looks like red brick, but in fact is a yellow brick with some kind of a red colour wash applied. What type of pointing is this, and how should we repair it?

A. Your building is *tuck pointed*, a common practice in the eighteenth and nineteenth centuries. Tuck pointing gave a coarse brick building a more refined appearance.

Coarser bricks, which were inconsistent in colour and shape and with large mortar joints, were camouflaged by using coloured mortars and sometimes washes, leaving a thin white bead (tuck) of lime mortar.

Tuck pointing is very much part of Georgian Dublin. While very few bricklayers can do this type of work today, some still can. You should try to find one of these skilled craftspeople, as tuck pointing is now rare.

Today, various forms of tuck pointing are being applied inappropriately to some brick buildings that show

Below left: Wigging, a Dublin style of tuck pointing, applied to a yellow clamp-fired brick. The example in the photo is old and weathered

Below right: New tuck pointing. North Great George's Street, Dublin

no sign of its previous use. It should only be applied to buildings where there is evidence that it once existed.

For a full explanation of tuck pointing and other pointing types, refer to the Practitioners section ('Brickwork', page 98).

CEMENT

Q. Is cement used in any mortar, plaster or render mixes discussed in this book?

A. No. Cement is not used in any of the mixes discussed here. As a general rule, it should not be used in the care and repair of old buildings. Cement has many applications in new-build where it may be used more appropriately.

CEMENT — WHY NOT?

Q. If they'd had cement in the past, I reckon they would have used it. So why not use it now on my old building?

A. Builders may indeed have used cement if it had been available. If they had, it is possible that their buildings would not have lasted as long as those built using lime. The evidence is everywhere to see. Not only are houses built 200–250 years ago still standing – they are very highly valued. They are likely to out-last most houses built today which have a much shorter life expectancy. Lime produces permeable, flexible structures. To maintain these positive attributes, repairs should be done with the same materials.

CEMENT — JUST TO BE SURE...

Q. My builder says he will only use natural hydraulic lime if he can add a small amount of cement to it. He wants to make sure it sets.

A. Your builder's experience of lime is probably limited to the common bagged (hydrated) variety of high calcium lime. It is used worldwide in sand and cement mixes to make them more workable. This hydrate is never used by itself, as it would be too weak. This presumably is what your builder understands. Natural hydraulic lime is not the same type of lime and should not have any cement added.

CEMENT RENDER – REMOVAL

Q. The thirty-year-old sand and cement external render on our gable wall was cracked, letting in rain. It had separated from its background, so we removed it. We can now see that the stone in the gable wall was built with a mortar that looks like mud. On removing the render, some of the mud mortar between the stones was removed too. What should we do now?

A. Mud was commonly and successfully used as a mortar. It was often protected with an external lime render to prevent wash-out by rain. You will have to point before rendering. This can be done using either mud or a weak lime mortar mix. The building can then be rendered using lime.

It is critical that you replace the small stones (pinnings) in larger mortar joints in walls such as these. Stone pinnings not only prevent these soft mortars from getting washed out on exposed stone – they also provide tightening to individual larger stones and therefore an element of overall structural stability to the wall face.

A multitude of stone pinnings can be seen in these eighteenth-century mortar joints. County Galway

CEMENT RENDER — RETAIN

Q. I've recently bought an old house that requires a number of repairs. There are many problems, including the slated roof. In the 1960s, the house was rendered externally using sand and cement, and this seems to be sound. However, I was told that I should remove this and replace it with a lime render as this will allow the walls to breathe and prevent future problems with damp. I am prepared to do this but would like to know the procedure and the appropriate lime mix to use.

A. Unless there are obvious problems with damp – such as damp patches on the internal face of the external wall, deteriorated internal plasterwork, rotten skirting boards etc. – leave it alone. Live in your old house for at least one year, experiencing all four seasons, before making any decisions. Check whether there are any obvious problems such as damp patches.

A cracked sand and cement render will allow water in, but prevent it from getting back out. Rising damp will appear at higher levels on the internal face of external walls because it cannot evaporate out through the sand and cement render. Removal of renders can create problems such as loosening of brick reveals around windows and doors. These may have to be rebuilt or partly rebuilt before applying the new render. If the render is in good condition, leave well enough alone and wait and see.

For now, concentrate on the roof structure and make sure that it is sound. Missing or damaged slates should be replaced. Make sure there are no leaks, and that there are adequate, clear-running gutters, downpipes and flashings, and that these are maintained regularly.

CEMENT POINTING

Q. My rubble stone outbuildings require pointing because of noticeable mortar loss. The most common type of pointing in my area is of the raised variety (proud of the wall face), and carried out in sand and cement. Is this the correct thing to do?

The original render has been removed inadvisedly, and the wall has been grit-blasted and strap-pointed using cement

Opposite: Lime dashing and limewashing following removal of cement pointing at Kildrought House, Celbridge

A. This pointing is called *strap pointing*. For your building, it is both the wrong style and the wrong material. Unfortunately, this is a very common and popular style of pointing. When it gives problems, it is extremely difficult to remove without damaging and staining the stone. It catches rainwater and prevents evaporation through mortar joints.

A lime mortar with a flush finish is preferable. The most common way of finishing this joint is by beating with a stiff brush to expose the aggregate in the sand. Mortars should always be weaker than their host material (brick and stone).

BUILDING LIMES FORUM

Q. How can I learn more about lime?

A. For further information, visit the websites of the Building Limes Forum Ireland and its affiliate in the UK. www.buildinglimesforumireland.com www.buildinglimesforum.org.uk

PART

2

PRACTITIONERS

INTRODUCTION

The building crafts which work with mortars, renders and plasters are known as the *wet trades*. There are three distinct groups of practitioners within the wet trades:

- Stonemasons
- Bricklayers
- Plasterers

Not so long ago, these mortars, renders and plasters were lime based – as they had been for over a thousand years in Ireland and for much longer in many other parts of Europe.

As the lime mortars used by each of the wet trades are often similar, a general introduction to lime mortar is useful here. Lime mortars that are more specific to an individual trade are covered under that trade.

Training is crucial for anyone about to work with lime, especially those who only have experience of working with cement. Incorrectly, limes are often seen as similar to cement, but they are different in selection, preparation, application, setting and long-term performance. Lime work also requires additional protection from the elements and increased after-care compared to cement.

Individual practitioners are advised to read the complete book rather than their own particular section only. This will provide them with a good general overview of working with lime.

LIME KILNS

James F. Collins, in his excellent book, '*Quickening the Earth*', *Soil Minding and Mending in Ireland*, explains how the burning of lime was very much a farming activity, widespread throughout Ireland, particularly from the seventeenth to the twentieth century. He describes two main types of traditional kilns – the running kiln and the standing kiln.

RUNNING KILN

The running kiln was called *tine reatha* in the Irish language. It was also called a perpetual, continuous, commercial, sale or draw kiln. This was a common type of kiln, and many still remain on the landscape.

The running kiln was built in stone, often limestone, and usually set into a hillside in order to retain heat and provide easy access to the top of the circular, open pot. This stone pot was commonly lined internally with mud, a well-worked yellow clay daub, laid on like a plaster in order to protect the stone. Into this pot were put alternate layers of broken limestone (fist size) and fuel such as turf, coal, culm (low-quality coal), firewood or charcoal or a mixture of fuels. The kiln was fired from the bottom. The process of converting limestone to quicklime could take about three days and nights. After this, the quicklime was drawn from the bottom of the kiln. As quicklime was drawn out, more fuel and limestone could be added to the top. It was

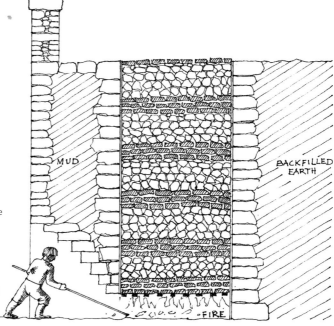

Running kiln. Conjectural drawing showing alternate layers of turf (peat) and limestone. The fire at the base is only temporary and is no longer needed once the alternate layers of turf take light. As quicklime is extracted at the base, additional turf and limestone are loaded at the top. This sequence of activity can continue over a prolonged period of time

possible to keep such a kiln burning and producing quicklime over a prolonged period of time, possibly months. There was a considerable saving in fuel as a result of the residual heat in the thick walls from continual firing.

STANDING KILN

The standing kiln was called *tine aoil seasta* in the Irish language. It was also called a French, flare or arch kiln. This type of kiln is more than likely older than the running kiln. It was much in use in the seventeenth and eighteenth centuries.

The fuel and limestone in the standing kiln were not mixed as in the running kiln but kept separate. An arch of limestone was built over a fire pit and on top of the arch was laid more limestone, up to the top of the kiln. The fire in the fire pit was kept going by raking out the ashes every so often and throwing in additional long-flame fuels such as turf, firewood or brushwood. About the time that the

Standing kiln. Conjectural drawing showing the limestone and fuel kept separately. This is a single burn kiln. Turf is thrown into the arch or arches and the fire at the base kept burning until the limestone is calcined to quicklime

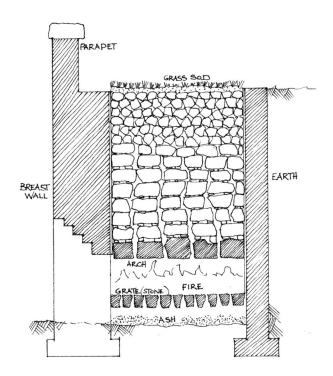

limestone had converted to quicklime, the arch had collapsed or was near collapse.

Larger pieces of limestone could be burnt using the standing kiln, and this was a labour-saving advantage. On the other hand, it used more fuel, and the residual heat in the walls was lost because only a single burn could be completed at any one time.

TEMPORARY KILNS

Temporary kilns were also used. They were constructed of stone and/or sods, or could be a vertical hole dug a short distance back from the face of an earth bank or peat bog. There is an account of one on Tory Island off the coast of County Donegal having been built with cut turf, chimney style, with limpet shells placed inside and the turf kiln itself set on fire. The fuel for temporary kilns could be turf, furze, scraws, bean straw or fern fronds.

A modern temporary kiln built as part of a one-day workshop. It is lined with mud and filled with alternate layers of coal, limestone and shells

TRADITIONAL LIME KILN CONSTRUCTION FOR FARMING: EARLY TWENTIETH CENTURY

Grants for the construction of new lime kilns from the Department of Agriculture were availed of by Irish farmers from the early 1930s to the late 1940s; grants for the repair of existing kilns were given from the early 1940s. A farmer availing of such a grant had to maintain the kiln and sell lime to local farmers at a price set by the Department of Agriculture.

The new kilns were to have a capacity of 300 cubic feet and to be built using stone that had a depth of 15–18 inches, laid in mortar. Walls were c. 5 feet thick and hearted with well-worked yellow clay daub. The internal face of the stone pot was lined with the same material.

In the 1930s and 1940s, 169 new kilns were built, while 54 were repaired in the 1940s. Although the grants were available in the early 1950s, they were not availed of.

Today, none of the kilns described is in operation on farms; crushed limestone rather than burnt lime is applied to the land with great success. A very small number of large, modern lime kilns produce sufficient lime for the needs of various industries, including building.

Below left: A small new kiln at Athenry, County Galway

Below right: Quicklime being extracted from a lime kiln at Narrow Water, County Down, for use in the conservation and repair of old buildings

SAFETY

Some procedures involved in working with lime are hazardous, particularly older techniques that require working with quicklime. Modern specialist suppliers who provide most of the industry's needs will advise on the care required when working with the more dangerous aspects associated with lime. They will also provide lime safety data sheets on request. *Protecting the eyes, hands and skin is essential.*

While some of the techniques described here are not in widespread use, they may be revisited by specialist practitioners, especially in conservation work. For this, training is required, and every care must be taken regarding personal safety, the safety of others, and property. Safety at work is covered under the relevant Health and Safety Acts and enforced by the Health and Safety Authority (HSA) in the Republic of Ireland, and the Health and Safety Executive for Northern Ireland (HSENI).

TYPES OF LIME MORTAR

HIGH CALCIUM LIME MORTAR AND THE LIME CYCLE

High calcium lime is produced by burning (*calcining*) pure or near pure limestone in a kiln. In the past, where limestone was less readily available, sea shells were also burnt. The majority of Irish limestones appear to have produced high

The principal building lime mortars, from left to right: quicklime, lime putty and natural hydraulic lime, each with sand

Top: Quicklime obtained from
burning Dublin calp limestone

Middle: Lime putty

Bottom: Mixing lime mortar,
County Mayo

calcium or feebly hydraulic limes. It was these limes that
were used for both agriculture and building. Hydraulic
limes will be discussed later.

When burnt to about 900°C, limestone changes to
quicklime (calcium oxide). In this process, the limestone
loses carbon dioxide gas which accounts for 45% of its
weight. The residue which is drawn out of the kiln after
burning is the quicklime. The quicklime is composed of
lumps about the same size as the stone that was placed in
the kiln (about fist size).

Quicklime was once transported to the place of work
by horse and cart, by boat or by train. On large projects –
such as the building of the twelfth-century castle at Trim,
County Meath – the kilns were constructed on-site.
Transporting quicklime was a dangerous activity. If it came
in contact with water, an *exothermic* reaction resulted – the
quicklime would heat up, causing it to spit and then
explode. Fire was always a possibility. Transporting
quicklime was always hazardous, particularly on rough seas
or in an open horse-drawn cart.

Once the quicklime arrived on site, it was processed in a
number of ways:

■ It was run to *lime putty* (calcium hydroxide) which
was then mixed with sand and most commonly
used for plastering (see 'Plastering', page 122). *or*

■ It was wetted with just enough water to make a
powder (calcium hydrate) and then mixed with
sand and water to make a mortar. *or*

■ It was used to make a hot lime mortar by mixing it
directly with a combination of sand and water (see
'Stonemasonry', page 58).

High calcium lime hardens through *carbonation* –
the re-absorption of carbon dioxide from the atmosphere.
As it cannot harden without access to carbon dioxide, it
will not set under ground successfully or under water
(hence it is also called *non-hydraulic lime*). High calcium lime
can therefore be stored indefinitely if it does not have
access to air. Even under the best conditions – mild, warm

Mixing hot lime mortar (quicklime, sand and water) – a dangerous activity

and slightly moist weather – it hardens slowly. In cold, wet weather, carbonation can cease and high calcium lime mixes can be susceptible to damage by frost and rain.

Carbonation occurs more successfully in high calcium lime mortars in the following conditions:

- The weather is reasonably mild. Late spring to early autumn is best.
- There is controlled water loss from the lime mortar, not too quick and not too slow.
- The water content of the mortar mix is kept as low as possible so that less water has to leave the mix.
- In the case of plasters and renders, individual coats are not applied too thickly.
- Surfaces are open, such as occurs from the beating of mortar joints with a stiff brush in the pointing process. The use of a wooden float in plastering

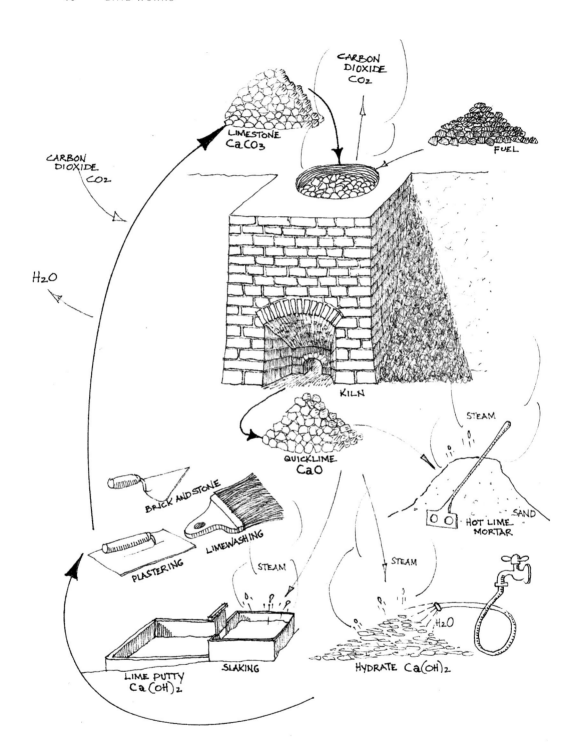

CARBON
DIOXIDE
CO₂

LIMESTONE
CaCO₃

FUEL

CARBON
DIOXIDE
CO₂

H₂O

KILN

STEAM

QUICKLIME
CaO

BRICK AND STONE

LIMEWASHING

HOT LIME
MORTAR

SAND

PLASTERING

STEAM

STEAM

STEAM

H₂O

LIME PUTTY
Ca(OH)₂

SLAKING

HYDRATE Ca(OH)₂

Opposite: The Lime Cycle showing how limestone or shells, after burning in a kiln, produce quicklime which is then used to create either a hot lime mix or a mortar using a hydrate or a lime putty. Once exposed to the air, the lime re-absorbs carbon dioxide, hardens, and returns to stone – a slow process

also leaves a more open surface than one finished with a steel trowel.

- Renders are thrown on rather than laid on with a steel trowel, as in the case of wet dash, giving a larger surface area to facilitate the absorption of carbon dioxide.
- Plaster or render backing coats are well scratched, allowing increased absorption by carbon dioxide.
- An occasional light spray of water is applied to ensure the mortar or coating does not dry too quickly. The carbonation process is assisted as a result.
- The joint sizes between stones are not excessive. If they are, they are pinned using smaller stones.
- The work is protected from rain, frost, wind and sun by wetted hessian or plastic sheets, or bubble wrap in cold weather. This is done during the course of the work and for an extended period of weeks afterwards.
- Allowance is made for slower setting times when scheduling work, compared to cement.

Carbonation is slowed down or does not occur at all in the following conditions:
- The weather is continually cold and the work is not protected.
- The weather is too wet and damp, with little or no drying.
- The weather is too warm, making the mortar dry too quickly if the work is not protected under damp hessian, supported nominally 100mm from the wall face, or lightly sprayed with water at regular intervals.
- The background wall is always wet or damp. The pores of the brick or stone and the mortar are saturated with water, preventing carbon dioxide from reaching below the surface.
- The work is enclosed too tightly in plastic sheeting or similar, not allowing sufficient air

movement to facilitate the absorption of carbon dioxide.

- The wall of brick or stone is too dry, has not been pre-wetted, and absorbs the water from the mortar mix too quickly.

POZZOLANIC LIME MORTARS

Since Roman times, *pozzolans* (volcanic ash or clay brick/tile dust) have been added to lime mortar to create a quicker set, one that is less reliant on carbon dioxide. On being heated and then combined with calcium hydroxide (lime putty), the minerals silica and/or alumina commonly found in clay (subsoil/mud) become reactive and create a *pozzolanic set.* With pozzolanic lime mortar, it was possible to build under ground and under water. A pozzolanic set allowed the production of concrete for building, something that was impossible if carbonation was relied upon solely because it would be too slow.

The Romans famously used clay tile/brick dust in their lime mortar for building Hadrian's Wall in Great Britain. It was used there to give a faster set and reduce the mortar's vulnerability to frost and rain.

In 1753, George Semple used pozzolanic mortars in the rebuilding of Essex Bridge on the River Liffey in Dublin. In his account of the work, he mentions the addition of brick dust to a lime mortar which was used hot for the foundations of the bridge.

In the past, every lime kiln also produced ash as a result of burning wood, turf or coal. This ash could be added to the mortar to create a pozzolanic set.

The charcoal created by burning wood is frequently found in medieval mortars here in Ireland. However, wood ash can be very unpredictable in the degree of pozzolanic activity it provides. This depends on both the species of wood being burned and its particular burning temperature.

Brick dust was often the result of soft brick fragments being thrown under the wheels of the mortar mill (*roller pan mixer*). These were crushed and mixed in with lime, sand

Left: Fine brick dust as a pozzolan about to be added to a lime putty and sand mortar mix

Right: Crushing the baked mud lining from a temporary kiln for use as a pozzolan

and water. Brick dust imparts a pozzolanic set only if it is very fine. Otherwise it behaves as an *aggregate*. The softer, lower or under-fired brick produced the better pozzolan compared with hard-burnt brick. Brick dust is available today as a pozzolan.

In times past, mud was used to line the inside of kiln pots in order to prolong its life from burning limestone. This burnt clay could have, intentionally or unintentionally, been drawn from the kiln with the quicklime. The result of burning and later crushing this clay would have been a pozzolan. This pozzolan, when added to quicklime, sand and water, produced a pozzolanic lime mortar.

NATURAL HYDRAULIC LIME MORTAR

Natural hydraulic lime also comes from burning limestone. In this case, it involves a limestone that is less pure than the type used to produce high calcium lime. Argillaceous (containing clay) limestone and/or siliceous (containing silica) limestone, when burnt, produce reactive aluminates and silicates that create a hydraulic set when combined with calcium hydroxide. A hydraulic lime can set in water and under ground. Dirty limestone such as Dublin's calp limestone is known to have produced variable hydraulic limes in the past.

Compared to high calcium lime mortar, hydraulic lime mortar sets faster and with increased compressive strength.

Today, compressive strength of hydraulic lime is assessed in Newtons per square millimetre at 28 days (N/mm^2 at 28 days).

The term 'natural' refers to a hydraulic lime to which no cement, pozzolans or other setting agents have been added. Natural hydraulic lime is imported into Ireland.

In the nineteenth century, hydraulic limes were assessed on their ability to set in water and were classed as *feeble*, *moderate* or *eminent*. Since modern NHL classifications do not correspond exactly to these designations, the following chart is only an approximation. In addition, modern limes may not have the same characteristics as old limes.

Nineteenth Century Classifications	Twenty-first Century Equivalent in N/mm^2 at 28 days
Feebly hydraulic	NHL1 (New addition)
Feeble to moderately hydraulic	NHL2 ($\geq$2N/mm2 to $\leq$7N/m₁
Moderate to eminently hydraulic	NHL3.5 ($\geq$3.5N to $\leq$10N/mm2
Eminently hydraulic to Natural Cement	NHL5 ($\geq$5N to $\leq$ 15N/mm2)

The compressive strength ranges for NHL2, NHL3.5 and NHL5 are so wide that there are great overlaps between them. Since 2007, more feebly hydraulic lime (NHL1) has become available commercially, although at the time of writing it is not included in the European standard.

Natural hydraulic limes used in new-build sometimes have air entrainment agents added to them to increase workability and reduce susceptibility to frost.

There is a current tendency to use the higher strengths of NHL in the repair of old buildings, even though the existing weaker mortars have exhibited an exemplary performance. While this may be due to the lack of experience or confidence in their use, the inappropriate use of excessive-strength mortars can give rise to problems.

Limestone flags being laid on a bed
of natural hydraulic lime mortar on
a limecrete floor

GAUGED LIME MORTAR

At times in the past, high calcium lime and the weaker
natural hydraulic limes were sometimes *gauged* with a faster
setting material such as a:

- pozzolan
- hydraulic lime
- natural cement
- gypsum plaster

Even today, the construction industry uses a gauged
mortar mix of cement, high calcium lime (as a hydrate)
and sand to build concrete block and brick. Combined
with the excellent workability of high calcium lime,
gauging achieves faster setting times and increased strength
in compression. Gauged lime mortars need to be agreed by
the architect, lime manufacturer and builder before
proceeding with the work to hand.

SANDS

Although sand makes up the bulk of most mortars, renders and plasters, it is one of the most neglected aspects. Modern sands generally have particle sizes no larger than 2 mm, while many sands found in older mortars and renders are more akin to light gravels.

The modern wet trades know well that a good sand should be *sharp, clean* and *well graded*. Modern plants have machinery to do this, although the grading of sand often falls short of the ideal. Sharp sands are often achieved by crushing; clean sands by washing; and graded sands by combining the different particle sizes within a sand in the correct proportion. Well-graded sands have about a 33% void or air content. From Roman times to the present, the replacement of this air with a binder has resulted in a generic mix of 1 binder to 3 sand (1:3).

Many of the particle shapes we see in old mortars are round and not sharp. While they generally do not conform to modern best practice, they have stood the test of time.

In the past, masons worked with what was available locally, making adjustments as necessary. Adjustments to sands were mostly concerned with workability. With sand that was too coarse, bordering on gravel, the larger particles had to be removed before it could be used as a mortar. Within living memory, sand was thrown by shovel against large mesh screens to remove the larger particles. Before that, the larger particles would have been picked out by hand. Poorly graded and coarse sands were improved by the addition of mud, a finer sand, or an increased amount of lime.

Today, because the grading of sand is not always what it should be, adjustments may still be necessary. By a simple field test, sand delivered to the site can be checked for air content (see 'Specifiers'). If the air content is found to be higher than acceptable (33%), additional lime may have to be added. Of course, there is a limit to how much extra lime can be added to make up for the poor grading of sand. If too much lime is added, it can result in shrinkage, extra cost, and either increased or reduced strength. Any alterations to specified mixes must be agreed by the architect.

Sand after grading. Starting on the left: B.S. sieves 5mm, 2.36mm, 1.18mm, 600 microns, 300 microns, 150 microns, and lastly a container for dust

WATER

- All water used should be drinkable.
- Excess water in a mortar mix will result in a loss of strength. This is due to voids remaining in the mortar after the water has evaporated.
- Sufficient water is essential to the carbonation process of high calcium lime mortars and the *chemical set* process which occurs in hydraulic lime mortars.
- If mortars dry out too quickly, they will not harden or set properly. Therefore all lime mortars need to be protected from drying out too quickly by using damp hessian or similar.
- The use of plastic sheeting to protect non-hydraulic lime mortars can prevent carbonation. Access to air should be provided.
- If there is a danger of frost, damp hessian should not be used. Instead, dry hessian and plastic bubble wrap may suffice, depending on the severity of the anticipated frost.

SOURING OUT

Souring out involves mixing high calcium lime, sand and water and then storing it for a period of time. This used to be considerable – six months or more. During storage, and to protect it from the effects of sun, wind, frost and rain, the lime mortar is placed under a sealed cover so it does not carbonate or dry out. This was common practice up to about fifty years ago but is now largely forgotten. The high calcium lime used could be in the form of a quicklime, lime putty or hydrate.

Where quicklime, sand and water were used, this souring-out time allowed the quicklime to break down further and combine with the sand. This reduced the risk of quicklime *inclusions* (small lumps) expanding and causing popping in plaster backing coats. Even rather poor sands produced good, workable mortars if the souring out method was used.

Up to the 1960s, most builders' yards and builders' providers stocked soured-out lime mortars for use with cement or gypsum. With the arrival of the easier-to-use liquid mortasisers, this fell out of favour. But for workability, soured-out lime mortar is still best. It has a place today in the replication of old mortar mixes for the conservation and repair of old buildings.

ADJUSTMENTS TO LIME MORTARS

In the past, there were local variations in both limes and sands. In the hands of skilled craftspeople, adjustments were made to lime mortars that may never be understood fully today.

Traditional adjustments would have involved:

- Adding a pozzolan to the mortar.
- Varying kiln temperatures (higher temperatures can increase hydraulicity in hydraulic lime).
- Adding clay to the kiln to create a pozzolan.
- Varying lime-to-sand content.
- Adding mud to the mortar.
- Varying water content (because of weather conditions or particular conditions of the job in hand etc.).
- Adding blood, urine, milk, salt, animal manure, hair, straw etc. to achieve a quicker/slower set, harder surface finish, longer life, better workability, reduction in cracking etc.

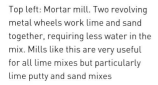

Top left: Mortar mill. Two revolving metal wheels work lime and sand together, requiring less water in the mix. Mills like this are very useful for all lime mixes but particularly lime putty and sand mixes

Top middle: A paddle mixer is another useful machine for mixing lime mortar

Top right: A modified rotary drum cement mixer with two mixing stones used to mix lime mortar

Right: A key indicator of high workability is the ability of lime mortar to stick to the underside of a trowel

Careful observation, research, scientific analysis, and the involvement of skilled and experienced craftspeople who are familiar with traditional methods, can result in a more dynamic approach to repair mortars on old buildings, rather than the more usual 'one or two mixes do all' approach.

Whether past or present, the main concerns of craftspeople are:

- Workability
- Setting times
- Appearance (particularly in plasterwork)

STONEMASONRY

INTRODUCTION

Ireland has two great traditions of working with stone — dry and mortared. While dry stone construction is the older by far, we are concerned with building stone in lime mortar. For the stonemason historically, hot lime mixes held many advantages and were used more frequently.

TRADITIONAL HOT LIME MORTARS

Hot lime mortars are very rarely used today, but were commonly used in the past. When quicklime, sand and water were combined, they produced heat. The quicklime

Lime mortar pointing in process at Ardfert Cathedral, County Kerry

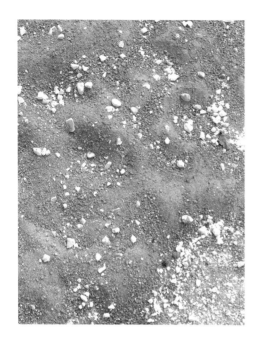

Above left: Quicklime and sand mixed together, dry. The quicklime has been reduced in size. Quite large core particles of unburnt limestone can be seen in the mix; these become part of the sand/ aggregate in the mortar mix. Water will now be added carefully to produce a hot lime mortar. This is a hazardous activity requiring personal protective equipment and training

Above right: A hot lime mortar (quicklime, sand and cold water) being mixed by shovel. Steam can be seen rising within minutes, indicating that the temperature is high and climbing. This is a traditional way of mixing stonemasonry lime mortars that were then used either hot or cold. Special training and safety precautions are required before attempting hot lime mixing

used would have been mainly high calcium. The resulting mortar was hot – very hot – with steam rising during mixing. Extreme care was essential while mixing, as a splash in the eye could cause blindness, and any contact with the skin would result in a burn.

Once the mortars were mixed, there were two choices – use the mortar while still hot, or sour it out and use it cold.

For the stonemason, there were a number of advantages in using the mortar while still hot.

- The mortar stiffened rapidly (within minutes) and work could continue without delay. In particular, the ability to continue building vertically was less restricted and the danger of collapse reduced in comparison with working with cold high calcium lime mortar.
- Wet stones could be laid without *swimming* (stones moving).
- Wet sand, a common occurrence, was never wet enough.
- Because of the stickiness/tackiness of hot lime mortar, very coarse sands in the mix did not create

a workability problem and were barely noticeable when working.

- Because of the heat, there was instant resistance to frost. Rapid evaporation also meant reduced water content in the mortar and gave further resistance. In the long term, hot lime mortars were more resistant to frost than lime putty mortars.

- Inclusions or small lumps of quicklime in the mortar mix expand when wet, causing expansion and swelling. As a result, shrinkage cracks closed, and the hot mortar filled minor gaps and holes that were left during construction. In general, the expansion was not severe enough to cause *jacking* (lifting and displacement of stones) but was sufficient enough to tighten up walls.

- On wall faces, full/swollen joints from hot lime mortar prevent excess ingress of water, plant growth etc.

- Hot lime mixes are possibly the best for workability. High workability translates to increased production and to full, solid and compressed mortar joints, particularly vertical joints which are often inadequately filled with less workable mortars.

- A good pace of work could be established in order to keep ahead of the hot lime mortar hardening. If the mortar was not used promptly, it would harden on the board. When this happened, more water would have to be added and the mortar then had to be used cold.

On the thick walls of defensive and other structures, it appears that hot lime mortars were semi-poured over the hearting stones in the centre of the walls, course by course, and probably hoed into vertical joints. The arch barrels of stone bridges were sometimes grouted over with hot lime mortars.

Hot lime mortars were used in renders and even in plaster backing coats, but not finish coats. Confusingly, what might appear to be the remains of a weathered render on stone is in fact sometimes the hot lime bedding mortar (of the stone) which has swollen and spread beyond the joint surface.

A high calcium hot lime mortar which has hardened after a few minutes has not carbonated – it has only lost water. Carbonation takes considerable time and occurs first at the outer faces of walls. The inner less-carbonated mortar is subsequently protected from being washed out by rain.

Pozzolans (brick dust etc.) were added to hot lime mortars to achieve a faster set, as well as in concretes, underground work and in water. The combination of heat, a pozzolanic set, and top-class workability created mortars that probably have never been bettered.

Lumps or inclusions of quicklime in old mortar are easily observed on many old buildings. These are an indication that either a hot lime mortar or a primitive dry hydrate was used.

In the nineteenth century – with increased industrialisation in the manufacture of products, stricter building codes, thinner walls, and the gradual erosion of local knowledge and ways of doing things – hot lime mortars became associated with bad practice.

SAND: COARSE

Q. I am unable to locate a coarse sand of 5 mm maximum particle size, graded down to dust for pointing an old stone

An old quartzite rubble wall built in lime mortar with a coarse sand

wall. The local sand, although right in every other way, is about half that size. Is there anything I can do?

A. This is a common problem. Sand suppliers generally stock sand which is suitable for the needs of new-build, so finding a coarser matching sand can be difficult. Use your local sand, but add some coarser material, if available locally. Sometimes 5 mm chippings/aggregate is available that contains not only 5 mm particles but also some finer grades. This is not a material that can be used as a sand by itself because it has an insufficient amount of *fines* (smaller particle sizes). Add a small percentage of this coarser material to your sand, say 5%. Mix with your lime binder, and point a sample panel and compare it with the original.

NATURAL HYDRAULIC LIME – USING

Q. What problems, if any, can I expect when working with natural hydraulic lime mortar? My only experience to date has been with Portland cement-based mortars.

A. There should be few problems. Natural hydraulic lime comes in a paper sack as a hydrate (powder) and is mixed relatively easily with sand and water.

- Thorough mixing is essential, and for a longer period of time than you are used to. Do not add extra water to accomplish quicker mixing and workability.
- The batching of NHL and sand should be done accurately and not measured with a shovel but with batching containers such as plastic buckets. Batching by weight is far more accurate. The water content should be controlled to the minimum for the job.
- If it is left for a short period when fully mixed, and then knocked up again, its workability usually improves.
- If possible, where workability is critical, a lower strength of NHL mixed with less sand is preferable to a higher strength of NHL mixed with additional sand. The same strength in compression can often be achieved by doing this.

Other factors may prevent this happening; check with the specification, architect and manufacturer.

■ The sand quality is very important. It should be well graded and clean. Coarser sands are generally less workable than finer sands.

■ Natural hydraulic lime sets more slowly than cement. You may have to leave sections of wall for a longer period before returning to add another *lift*.

■ A slower set means that natural hydraulic lime is more susceptible to frost, wind, rain and sun than cement. Allow for a longer period of protection and spray occasionally with a fine water spray mist.

■ Consider attending a training workshop where you will learn both the theory and practice of working with lime.

NATURAL HYDRAULIC LIME – SELECTION

Q. Why should I not use the strongest natural hydraulic lime to achieve faster setting and resistance to frost?

A. Where it is not warranted, increased hydraulicity can create increased dampness and deterioration of masonry units. Increased mortar strength will reduce flexibility within joints, possibly leading to cracking of masonry units. Long term, it will make the recycling of masonry units more difficult. Where lime mortar is used in conservation work, it should generally be compatible with whatever lime mortar already exists.

POINTING STYLES – RIGHT AND WRONG

Q. There are right and wrong styles of pointing stone. What are these?

A. The Right Pointing Style
■ A historic structure may require the replication of an existing joint finish.

- For most work, a *flush mortar joint* is the most appropriate. It should be beaten with a stiff brush to finish.
- Although a flush mortar joint may be appropriate for most situations, if the stones are rounded and reasonably hard, the mortar can generally be kept back from the face of the wall – *i.e.* recessed and not flush to the face. This prevents the joints from looking exaggerated in size. It also prevents *feather edging,* where mortar is taken to a thin edge over rounded stones, allowing water to enter from behind. Flush pointing also reduces the visual size of the rounded stones in the wall. Flush pointing of rounded stones is less common in Ireland than in some other European countries such as France and Spain. Recessed joints are finished by beating with a brush. The size of the brush may have to be adjusted to accomplish this. If the wall is to be rendered or plastered, there is a case for flush pointing the rounded stones. This will create a reasonably flat substrate to receive the first coat of plaster or render. Otherwise, it is normally best to slightly recess joints as described, with hard,

The right pointing style. Flush lime mortar joints beaten with a stiff brush to expose the aggregate in the sand

rounded stone. The re-insertion of stone pinnings in the joints of rounded stones may be critical to prevent movement and possible collapse. They also reduce visual joint size and allow the smaller exposed amounts of mortar to carbonate more effectively.

Wrong Pointing Styles

- *Weather struck pointing.* This takes the form of a sloping joint with the intention of throwing water off the wall. It is usually carried out in sand and cement mortar.
- *Strap pointing.* A projecting joint with a flat face, consistent width and parallel sides, usually only seen in sand and cement mortar.

Roughly executed weather struck pointing in sand and cement inappropriately applied to a nineteenth-century limestone wall

Strap pointing in sand and cement inappropriately applied to a nineteenth-century building

A late medieval rubble limestone structure with evidence of previous pointing repairs, holes and missing stone pinnings. A minimum intervention approach is required, repointing only where absolutely necessary

- *Jointed with a round steel bar.* This is a 1960s' development in brickwork which gives a compressed joint but is not appropriate to stonework.
- *Recessed joints.* Joints are raked out to a uniform depth throughout and left. This is a modern concept of emphasising the stone – it catches water and is not suitable for old work.

POINTING RUBBLE STONE BUILDINGS

Q. I have been asked to use lime mortars to point a number of rubble stone buildings on a nineteenth-century estate. All joints are to be finished flush, with pinning stones inserted, where missing, in large joints. The conservation architect has decided that only walls showing serious mortar loss are to be pointed. Matching coarse sands and the type of lime and mix ratios have been chosen. A roller pan (mortar mill) mixer will be on site.

I have only worked at building modern walls previously, and these have been pointed with sand and cement. Can you take me through the stages of how to point rubble stone buildings using a lime mortar?

A. Pointing is carried out in the following stages:
1. Raking out
2. Washing out/Brushing down
3. Mortar Mixing
4. Dampening
5. Mortar Application
6. Pinning Insertion
7. Galleting
8. Beating/Finishing
9. Protection and Curing

1. **Raking out**

Start at the top of the structure and rake out existing mortar joints to the agreed depth. This can be done in scaffold lifts, with each lift being inspected before pointing commences, or by complete individual wall faces or buildings. The depth of the rake-out should generally be greater than the width of the mortar joint – from a minimum of about 25 mm up to twice the width of the mortar joint. Common sense is needed here.

The raking out should be sharp and squared off, not 'V' shaped. The *arrises* (edges and corners) of the stones

A plugging chisel and a lump hammer being used to rake out mortar joints in preparation for pointing. Other specialist chisels for raking out are available

should not be damaged; in most cases, a plugging chisel used with a lump hammer is adequate.

Non-ferrous chisels are available for situations where the transfer of steel from the chisel to the stone could result in staining by rust. Chisels with tungsten tips are sometimes used. Angle grinders should not be used because they widen joints and damage stone.

Sometimes, mortar that has been raked out is used again as an aggregate in the replacement mortar, more so with historical structures. Check with your conservation adviser.

2. **Washing out/Brushing down**

Some careful washing out under pressure may be necessary. When washing out, care must be taken not to remove excessive amounts of soft mortar, or the face stones of the wall may become dislodged. The amount of water used should be kept to the minimum.

If the joints have been washed down already, there will be little need for brushing. Otherwise, brush down starting at the top of the building in order to clean out debris from the joints. The brush should be soft, with bristles sufficiently long to reach the back of the mortar joints.

3. **Mortar Mixing**

The lime type, the sand and the mix ratio will be specified by the conservation adviser. Generally, mortar mixes for pointing rubble stone buildings are either a lime putty, sand and pozzolan mix, or a natural hydraulic lime and sand mix.

Lime putty is generally more difficult to mix with sand, but the roller mill will make this easier. As this type of lime usually contains sufficient water, it may not be necessary to add more to the mix.

(Lime putty, pozzolans and natural hydraulic limes are discussed elsewhere in this book.)

The mortar mix should be stiff but workable. Excess water produces a weaker, brighter-coloured

A stiff lime mortar with a low water content but good workability, suitable for pointing

mortar. Keep the sand dry to control the amount of water in the mix. It is essential to spend additional time mixing rather than adding more water. Pointing requires a stiff, plastic mortar that can be lifted off the pointing hawk with a flat jointer. An experienced stonemason will make minute adjustments to the mortar by occasionally knocking up, turning over, and beating with a trowel.

Mix ratios need to be clear, and measuring devices (batching boxes, buckets or containers) should be agreed before work commences. The more simple the procedure, the more likely it is to succeed.

Any mix ratio adjustments should be agreed between the architect and the contractor. Insufficient mixing time can sometimes be the problem. In general, lime mortars require a longer mixing time than cement mortars.

A hand-held water spray bottle may be all that is necessary to dampen the joints prior to pointing

4. **Dampening**

In advance of pointing, dampen an area of wall which is sufficient for about an hour's pointing work. How much water should be used in the dampening process depends on the weather conditions, the porosity of the stone, and the joint size and existing mortar in the wall. Concentrate on dampening the mortar joints rather than the stone. If the wall is too wet, leave it until it dries to being just sufficiently damp. In warm, dry conditions, frequent dampening down and control of water loss from the wall using damp hessian and plastic sheeting for cover may be necessary. In such cases, consider working on naturally shaded sides of the building throughout the course of the day.

5. **Mortar Application**

Pointing trowels are not suitable for placing mortar at depth and with compression. They also naturally leave the wrong style of finish. Instead, *flat jointers* should be used to insert mortar off a pointing hawk into both horizontal and vertical joints.

Flat jointers are variously called jointers, flat bar jointers, flat irons and jointing irons. They range in size from 3mm upwards and are made of flat steel. Non-ferrous versions are available if required.

A pointing hawk has a flat top and a handle. The flat top measures approximately 150 mm x 150 mm but can be larger. It can be made of wood, plastic or metal.

Right: The lime mortar is lifted off the pointing hawk with the flat jointer

Below: Tools for pointing. Bottom left shows a plugging chisel with brushes above. The tools on the right are flat jointers of varying width, shape and size for the application of mortar

Above left: For filling vertical joints, the mortar must be sufficiently workable to stick to the flat jointer

Above right: The pointing hawk is held close to the horizontal joint and mortar is scraped off the hawk into the joint without staining the stone. A right-handed person works from right to left, and a left-handed person from left to right

A rammer is used when mortar must be compressed a considerable distance in from the face of the wall. This rammer is made from a length of flat mild steel bar welded to a round reinforcing bar

A rammer made from a length of flat mild steel bar of suitable thickness and width to fit the tighter mortar joints

For horizontal joints, hold the edge of the hawk close to the wall so that the mortar can simply be pushed into the joint, taking care not to stain the stonework. For horizontal joints, always work from right to left if you are right-handed, and vice versa if you are left-handed. This ensures that the mortar is always pulled back to an anchor. Fill the joints full, or slightly more than full, leaving no holes, cracks or depressions in the mortar joints.

Don't over-work the mortar with the flat jointer. Over-working brings the water in the mix to the surface. Insert the mortar in the joint using the least amount of moves. Increased water at the surface of the mortar joint results in a lighter colour finish and slumping of the joint. This leads to the mortar running down the face of the stone and a slow-down in hardening, thus delaying the finishing stage.

Fill vertical joints, first by taking the mortar off the hawk onto the flat jointer, then applying it to the open joint. The mortar must be sufficiently workable to stick to the flat jointer. If not, it is very difficult to fill vertical joints, and time and materials will be wasted in the process.

The surface of the inserted mortar is sometimes dragged with the edge of the flat jointer immediately after insertion. This increases the surface area of the mortar joint, reduces the possibility of slump in wider joints, and slightly increases evaporation and therefore surface hardening.

Above left: Dampened limestone pinning being inserted into a full mortar joint during the repair of a wall beneath a granite window cill

Above middle: Hand pressure may be sufficient when inserting pinnings

Above right: Pinnings may need to be driven home lightly with a hammer

Very deep, open joints will have to be filled and compacted with a special tool called a *rammer* or *ramming iron/bar*. This has a long handle and a piece of flat steel fixed at right angles to the handle on the end.

6. **Pinning Insertion**

Pinnings are generally rather small stones, often flat on their beds and sometimes diminishing in size along their length to make them easier to insert into the mortar joint. If required, pinnings are inserted in both horizontal and vertical joints. Pinnings were commonly used in all types of lime-mortared rubble stone. It is only with the advent of cement that they are being used less today.

In traditional lime-mortared work, pinnings were used to:

- Reduce larger joint sizes and prevent mortar getting washed out.
- Prevent larger surrounding stones from dipping/tilting and becoming loose, resulting in possible collapse.
- Reduce the visual joint size, giving the work a more controlled and aesthetically pleasing appearance where work was not later rendered or plastered.

The pattern of pinning and the stone to be used should be discussed and agreed in advance by the professional adviser and the contractor. This will be based on observation of any pinnings still evident in the structure.

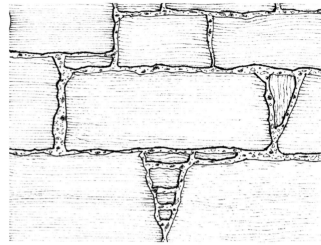

Above left: Pinnings inserted into the joints in rubble masonry prior to the joints being finished

Above right: Pinnings may be either natural bedded (centre bottom) or edge bedded (middle, right). The small stone (top, left of centre) is a sneck and not a pinning

Old photographs, if they exist, are always helpful.

Pinning stones which are loose but still in position can be removed during the raking out process. They can then be reinserted dry in the same position, ready for reinstatement when the wall is being pointed.

Some pinnings are often found on the ground where they fell many years ago. New pinning stones should match the geology, colour, size and shape of the existing ones. The style and pattern of pinnings should be the same.

Pinnings are best inserted:

- With their length (where possible) running into the joint.
- Laid on their *natural beds* (sedimentary and some metamorphic stones) in horizontal joints and tilted slightly downwards and outwards to throw off rain.
- In vertical joints, they are most commonly *edge bedded* to suit the joint shape rather than on their natural bed. This often varies, so follow the existing pattern.

Before inserting the pinnings, the joint should be full of mortar. Otherwise, it will be difficult to insert mortar later. Larger pinnings are best buttered with mortar before inserting them into the full joint. This ensures that there are no gaps/holes remaining on insertion.

Depending on the weather and materials being used, pinnings may have to be pre-soaked and allowed to dry until just damp.

When the pinnings are knocked into position with light blows from a hammer or pressed in by hand, they will cause a slight compression crack in the mortar around the perimeter of the pinning. The mortar needs to be re-compressed with a flat jointer and filled further before proceeding. Pinnings are generally inserted flush to the face of the wall and not set back or left projecting or set in out of balance at awkward angles.

Pinnings generally should have a flat face that is *out of twist* and runs with the face of the wall.

Pinnings which are porous – such as broken terracotta tile, pieces of brick and some sandstones – produce a faster set in most lime mortars. These are sometimes used, even though there is no precedent, where walls are to be plastered or rendered later, so the pinnings will not be visible.

7. **Galleting**

Gallets are sometimes confused with pinnings. Gallets are small stones, often of similar size, colour and

Galleting is usually only associated with eighteenth or nineteenth-century construction. This example is from County Limerick

The joints have been fully filled and
the mortar is now allowed to stiffen
before beating

The joints have been fully filled and the mortar is now allowed to stiffen before beating

shape, which are inserted into mortar joints to shallow depths. Gallets initially stiffen up surface mortar and prevent mortar-runs down the face of the masonry. They also protect mortar joints from getting washed out. While they are decorative, they do not serve a structural function as the larger pinnings do.

8. Beating/Finishing

It is not possible to begin this stage if mortar joints have not been filled fully in the mortar application stage. Holes and depressions will show, and attempts to fill them with fresh mortar just before beating will not produce a professional finish.

The time between the application of the mortar and the finishing stage varies with the lime type used, the porosity of the stone and the weather. The application and the finishing may be done on the same day, or there may be a gap of a number of days between each of them. The surface hardness of the mortar determines the optimum time for beating. This is explained below.

Before beating, close all visible shrinkage cracks in mortar joints with flat jointers. Although the mortar will be quite hard, there should be just enough plasticity remaining to accomplish this.

The mortar is now beaten with a stiff brush. (A nylon or natural fibre churn brush with a handle is particularly useful, although a thinner brush is

The mortar joints were beaten a little too early here. In another hour or so, the mortar would have been less plastic and shown the aggregate more clearly. In some cases, however, a preliminary beating like this allows for increased manipulation of the mortar to close cracks etc. and makes the final finishing process quicker

sometimes warranted for smaller joints.) Beating the lime mortar joint exposes the aggregate in the sand. A sand with large aggregate sizes will be very noticeable, showing the size, colour and shape of the particles. A mortar that is beaten while too fresh will come off on the brush, resulting in a loss of mortar, a stippled surface finish, and walling stones covered in lime. The sand particles in the mortar will also be hidden behind a coating of lime and not exposed as they should be.

If the mortar is not quite hard enough, an initial light beating will roughen and increase the surface area of the joints, allowing increased evaporation to take place and therefore slightly quicker drying. Shortly afterwards, the joints can be beaten again to achieve the desired finish. The initial beating procedure, although mostly face-on or at right angles to the wall, may also involve manipulation of the brush at various angles. This will close any minor shrinkage and compression cracks. Small amounts of mortar will also migrate from areas of slight excess to areas of slight shortage. This depends on the plasticity of the mortar.

It will not be possible to finish using the brush if mortar joints are left too long before beating. They will have to be scraped with a small steel blade.

At the finishing stage, any poor workmanship will be evident: marks of steel tools on mortar joints; long brush marks from brushing rather than beating; mortar running down the face of the masonry; untouched joints; mortar joints not filled out to the face; holes;

The final beating of the mortar
exposes the aggregate. Any lime
stains on the stone are removed
with a damp sponge

and *snots* (raggedy mortar projecting and overhanging past the face of the wall).

Finishing techniques vary from beating, to brushing and beating with a brush, to scraping with thin metal strips. There are advocates for these and other methods, but they all result in an exposed aggregate finish on a flush or near-flush joint that is compacted and full without cracks or holes. Any mortar stains on stone faces should be cleaned off with a semi-wet sponge.

9. **Protection and Curing**

If the work is not protected from the sun, wind, rain and frost, all the best efforts may be undone. Mortar should dry slowly in order to achieve its optimum strength. Slow-drying mortars are less likely to shrink and crack. Fast-drying mortars can turn to powder, crack, and dry out to a lighter colour than normal. All lime mortars require a longer period of protection than Portland cement mortars.

Drape hessian over the work at a distance from the wall of about 100 mm and keep the hessian wet. This ensures a high relative humidity in the mortar without the risk of drawing the lime to the surface or having mortar runs down the masonry due to over-zealous application of water to fresh mortar. Retain this environment for a number of weeks in normal weather conditions.

Scaffolding enclosed with
galvanised roof sheets and
monoflex sheeting to allow
all-weather working of
natural hydraulic lime mortars.
County Cork

Protection from frost is more problematic, requiring insulation and possibly low levels of heat. It is best to avoid working outside with slow-setting lime mortars from late autumn to early spring. In some instances, it may be necessary to provide a sealed cocoon of reinforced vertical plastic sheeting or even double bubble-wrap plastic sheeting, sealed at all joints with an insulated roof, and with a low level of heat. This, however, will prove to be expensive.

POINTING – ASHLAR STONE

Q. How can I point ashlar stone, with joint sizes averaging 3 mm, using a lime mortar?

A. The following will help you to point ashlar stone.

- Unless there is serious ingress of water or structural reasons that necessitate the pointing of ashlar stones, it is best to avoid pointing at all. Not only is it difficult, but there is also the risk of causing damage and staining to the arrises of the stone.
- Rake out as necessary. Don't use standard wedge-shaped chisels. Don't use a standard plugging chisel, as it is too thick and will damage the arrises of the stone. An industrial hacksaw type blade with a wooden handle may work, as will very thin chisels that display little or no wedging in their

A cut stone door surround in limestone displaying mortar loss from very fine joints. It is essential that the right tools, materials and techniques are used when repairing very fine joints

A fine hacksaw blade being used to remove decayed mortar loose from a stone capital in preparation for repointing

profile shape. Under no circumstances use an angle grinder to open or widen joints.

- If the existing mortar is flush, leave well enough alone.
- Vertical joints are tight at the face of ashlar stones, but often increase in width further in from the face of the wall.
- Clean, dry, compressed air will sometimes remove dust and debris from joints, but take care not to damage stone. Wear personal protective equipment to avoid an accident.
- Pre-wet the mortar joints until damp. A relatively small amount of water is required for ashlar joints. A plastic spray bottle may be sufficient. Because the joints are so small near the face of the stone, there is a danger of the mortar drying out too fast and failing. Therefore protection from the elements will be required to allow the mortar joints to set slowly.

- Sand with a maximum particle size of one-third the joint size – in this case, 1 mm – should be ideal. The sand should be graded from 1mm down to dust. Be careful to use a well-graded sand, not a single particle sized grit. Normal sand is too coarse for such small joints. The colour of the sand is important when matching existing mortar joints.

- Hair is sometimes added to the pointing mortar for ashlar stone. It is particularly useful when applying mortar to joints that are tight at the face but widen further into the wall. One method is to use waxed twine inserted into vertical joints where the joint widens at the back. This acts as a stop and creates a uniform depth of mortar insertion. This method is only used where grouting the wall with a lime mortar follows the pointing. Otherwise, sufficient mortar should be inserted to fill the joint and the immediate void behind.

- Open joints can be covered with masking tape and split with a sharp knife before the insertion of the

The masking tape (top left-hand corner) has been applied over a vertical joint and split with a sharp knife. Mortar can now be inserted without staining the stone, after which the masking tape is removed. County Cork

Old red sandstone and near-white carboniferous limestone pointed with a natural hydraulic lime mortar. The Old Waterworks, Lee Road, Cork

mortar. This protects the stone from being stained by mortar and also speeds up the work.

Apply the mortar off the hawk using a flat jointer to compress the mortar into the joint. The jointer should be very narrow – the edge of the jointer sometimes works best. A narrow kitchen knife with its end squared off may also be used to compress mortar into the joints. Plasterers' *small tools* which are used for modelling decorative plasterwork are thin bladed and come in a variety of shapes and sizes. These may also be useful for mortar insertion.

- Applying the mortar by compressed air is difficult because the fine nozzle of the gun gets blocked easily. Similarly, veterinary syringes become blocked easily, even when using very fine stone dust as an aggregate.
- Scrape off mortar flush to the masking tape.
- Protect the work from drying out by using damp hessian, plastic sheeting etc. An intermittent fine mist spray of water should be applied to ensure the mortar does not dry too quickly and fail.

Tamping/beating mortar joints with a small stiff brush

- Remove masking tape when mortar is relatively hard.
- Tamp mortar joints with a small, stiff brush. This gives a uniform textured finish. Sponge off any staining on the stone. Continue protecting the work with hessian or plastic sheeting. Use a fine mist spray of water to ensure the mortar remains damp until set.

POINTING – ASHLAR STONE – DIFFICULTIES

Q. We are pointing a public building constructed of ashlar stone. The joints (3 mm) are so small that it is impossible to insert the mortar between them. The latest suggestion is to widen the joints using an angle grinder. Is this our only option?

A. The widening of joints with angle grinders and power tools *should not be considered in any circumstances.* This will:

- Widen joints up to 12 mm from their original 3 mm dimension.
- Create cuts that over-run into adjacent stones, particularly at perpendicular joints.
- Turn a refined, well-finished structure, with accurately cut stone as part of its aesthetic, into a much coarser structure.
- Create wave-like cuts on horizontal beds because of the difficulty in controlling angle grinders and other high-speed circular blade cutting tools in that position.

Consider the sand size you are using. If it is standard building sand, the particle size will be too large. The maximum particle size needed is about one-third the joint size. The tools of application must be small enough to push mortar into these joints. (See 'Pointing Ashlar Stones', page 79)

HEARTING

Q. What is hearting? How should it be done using lime mortar?

A. *Hearting* is the stone and lime mortar carefully laid in the centre or core of the wall. Hearting makes up the difference, or fills the space, between the stones on either face of the wall. It is often referred to as *core fill*, but this gives the impression that its only purpose is to fill the centre of the wall and that the way in which this is done is not important. The correct method and sequence of laying hearting is critical to the structural stability and weather-protective function of the wall. Very thick walls on defensive structures are composed mostly of hearting.

The following description is for walls about 600 mm in thickness. Modification is needed for walls of greater thickness. It is assumed that one course of stone, with their lengths running into the thickness of the wall, has been laid on either face of the wall and that the centre of the wall is now ready to be hearted.

- Before commencing the hearting process, push small stones into lime mortar beds under face stones (working from the inside) to ensure the stability of the face stones. On walls of average thickness, this is best done by the stonemason on the opposite side of the wall because gaps and large bed joints are seen more easily from this position.
- Begin hearting by laying a bed of mortar in the centre of the wall.
- Butter the vertical back sides of all surrounding stones with mortar.
- The largest hearting stones should be laid first. They should not exceed the stones on either face

This drawing shows in plan the quoin stones at both ends of the wall and a through stone in the centre, spanning from one side of the wall to the other. The hearting stones in the centre of the wall and the face stones are laid transversely for structural strength

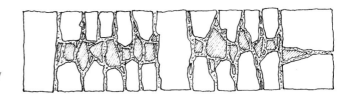

Left: Bed laid and vertical joints buttered with mortar in preparation for laying the hearting

Middle: Once the large hearting stones are laid, the medium stones, followed by the small hearting stones, are laid

Right: Hearting stones are slightly tilted towards the external face of the wall to throw off rain water. If using a hammer, take care not to push stones at the face of the wall out of position

of the wall in height. If they do, they will interfere with the next course of face stones.

- The hearting stones are usually those that are not of a suitable shape for use on the faces of the wall.
- The large hearting stones should be laid transverse to the face of the wall, as far as possible. In other words, hearting stones should be laid with their length running across the width of the wall. This strengthens the wall transversely. Longitudinal hearting can be desirable over openings or anywhere strength is required in the length of the wall.
- Hearting stones should not be forced into position, as this will cause the face stones to move outwards. The neater the fit, the stronger the wall.
- Medium-sized hearting stones should be laid next into the remaining spaces that are pre-buttered on all sides with lime mortar.
- Finally, the smallest hearting stones should be laid into full mortar joints and pushed into position. These are inserted not only into vertical joints between hearting but also between the vertical joints of the face stones.
- As far as possible, all hearting stones should tilt slightly outwards to throw off rain water. Hearting stones should not tilt inwards. While it is not always possible to accomplish this fully, it should be followed as a general rule.
- Through stones that tie opposite sides of the wall together should be laid at a slight fall outwards.

In walls thicker than 600 mm, through stones are rarely seen because the necessary length of stone is not available. Instead, bond stones are used that travel a considerable distance into the thickness of the wall. These should be tilted slightly outwards too.

- All joints between face stones and hearting on the exposed top bed of the wall should now be compressed with a flat steel jointer and further filled with mortar. If more small stones can be inserted into mortar joints at this stage, so much the better.
- Brush any hardened mortar off the top of the wall in preparation for the next course of stone.
- *Under no circumstances* should hearting be laid as dry stone, one upon the next. Stones should not be thrown or shovelled in, and all stones must be placed by hand. Sand, gravel, soil, mortar without stone or concrete should not be used to fill the centres of walls.
- If work should cease for the day at this point, and if there is a threat of rain, the wall top should be protected with damp hessian and plastic sheeting. Lay the plastic sheeting on the wall top at an angle to throw off rainwater. Insulated frames may be necessary for winter work.

NEW STONE-FACED HOUSE AND NATURAL HYDRAULIC LIME

Q. As a team of stonemasons, we have been asked to use a natural hydraulic lime 3.5 mortar to build a new stone-faced house. The stone will be 225 mm thick against a newly built insulated concrete block cavity wall of the house. A local limestone, black-to-dark-grey in colour, is to be used. The architect wants the stonework to have an authentic, traditional look and is not happy with the newly built houses close by which are faced with stone. We have to build a sample panel next week using natural hydraulic lime and are wondering what we should look out for.

Detail of an outer leaf of stone built against an insulated concrete block cavity wall using a natural hydraulic lime and a coarse sand. County Limerick

A. A number of key points will help you achieve your goal.

- A coarse sand ranging from particles of 5 mm graded down to dust looks good.
- Finish the joints by beating with a stiff brush (see 'Pointing Rubble Stone Buildings', page 66). This may have to be done the day after building.
- Gauge the quantities of NHL3.5 and sand accurately by using a batching box or measuring bucket.
- NHL mortars need to be mixed for longer than cement mortars. At least twice as long is usual.
- NHL mortars that are let stand for a couple of hours and then knocked up have increased workability.
- Control the amount of water in the mix. Too much water in a mortar used to build hard carboniferous limestone will lead to runs of mortar down the face of the stones.

New work showing stone laid against the outer leaf of a concrete block insulated cavity wall, hearted fully with stone and lime mortar. Wall ties are required from the stone to the block and across the cavity

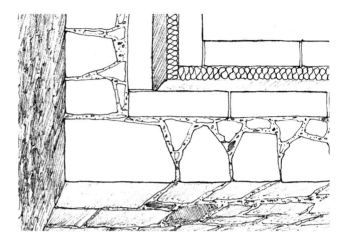

- Natural hydraulic lime will take longer to set than cement mortars, so allow for this. Also allow for the protection of the work from the elements for a longer period of time.
- Depending on the weather, protect the work with damp hessian to prevent it from drying out too quickly. This should be maintained for about two weeks.
- Lay the stone on its natural bed, not face bedded.
- As far as possible, lay stones with their lengths running into the thickness of the wall.
- Break all vertical joints and never have more than two stones on top of each other forming a straight vertical joint.
- Lay all stones to achieve a visual balance, not at awkward angles.
- Control joint sizes by selecting stones and/or by cutting with a hammer. A pitcher and a punch will achieve tighter joints, but this is more time consuming and may not be warranted. The architect should be consulted on this.
- To reduce joint size and to make the average joint sizes look more controlled visually, stone pinnings may be required in joints. They will tighten up larger stones and achieve a traditional look. Check with the architect before using.

Poorly constructed modern masonry wall built in sand and cement. Most stones are face bedded, out of balance, and have vertical running joints

- Heart between the back of the stone and the concrete block with lime mortar and stone, not just mortar or dry stone.
- Ensure that all wall ties are properly fixed in mortar bed joints.
- Good quality (large, square, long) stones of regular shape are required as *quoins* (corner stones) at corners and at openings.
- Ensure that vertical damp-proof courses at window and door openings are properly detailed before commencing.
- Follow good practice as set out in the architect's specification and have due regard for safety in everything you do.

DRY LOOK – NEW STONE

Q. I am building a new house faced with stone. It is a concrete block cavity wall construction faced with 225 mm of stone. The stone will have a dry look – no bedding mortar will be used in either horizontal or vertical joints between the stone. The stone will be tied back to the concrete block with expanded metal, and the walls will be approximately

Modern dry look wall built against a concrete block cavity wall of a house

5 metres high. Should I use a natural hydraulic lime mortar or a limecrete to fill between the back of the stone and the outer concrete block leaf in order to let the wall breathe?

A. Although the dry-stone look has become increasingly common, and your suggestion of using no bedding mortar between the stones on the face of a building is not unprecedented, it is potentially dangerous and may result in the wall collapsing. The reason for this is that the wall ties need to be bedded and secured in horizontal mortar joints between stones where they will resist tensile forces and prevent collapse. A dry look can still be achieved by keeping the mortar slightly back from the face of the wall in both horizontal and vertical joints.

All stones should be laid in mortar and tied back to the blockwork with appropriate wall ties. Expanded metal ties are often bent upwards or downwards to suit the irregular coursing of stone. Although this makes the work easier, it also means the stone wall can pull away from the block wall before the expanded metal tie comes under tension.

Because this style of wall is a hybrid of traditional and new, it is less important for the wall to breathe than it is in traditional solid wall construction. The benefits of using a lime mortar here are increased flexibility and a traditional appearance.

REBUILDING AN OLD GATE PIER

Q. What mortar should I use to rebuild an old gate pier? The stone is Irish limestone in large squared blocks, with quite small mortar joints of about 6 mm. The wrought iron gate also needs to be rehung.

A. Gate piers have to withstand many forces. A hanging gate exerts forces of tension and compression as well as shear from twist. If hung correctly, these forces are greatly reduced.

Building piers and hanging gates are highly skilled activities, particularly for heavier gates. Mortar needs to have tensile properties to stop stones moving or being pulled apart. A high calcium lime or the weaker natural

A hanging eye reset in a specially cut stone in an old pier. Molten lead will be used to fix the hanging eye (a dangerous activity) and the pier will be built higher using natural hydraulic lime mortar. County Limerick

hydraulic limes or pozzolanic lime mortars are unlikely to be successful. An NHL3.5 or even NHL5 lime and sand mix may suit.

The joint finish and sand can be matched to the remaining pier, but sand with a maximum particle size of 2 mm will suit 6 mm joints.

Gates are hung with a hanging eye from the pier near the top of the gate stile and a spud stone at the base of the stile. The traditional wrought iron hanging eye is fixed in the bed of a large stone in the pier, using lead. The spud stone (often granite) has a cast iron insert set in lead to receive the bottom of the stile. All the weight is transferred to the spud stone, while the hanging eye prevents the gate from pulling away from the pier. A properly hung gate, even though very heavy, will swing easily and exert little or no pressure on a suitably sized and well-built stone pier.

LIME MIXES – WEATHER

Q. We are working on an old stone structure. Now that winter has arrived, should we use an NHL3.5 mortar instead of an NHL2 mortar to get a faster set and avoid frost?

Above left: Natural hydraulic lime and sand batched by volume using a bucket. In this case, the mix is one part natural hydraulic lime to two parts sand

Above right: Lime and sand mixed, ready to add water

A. Presumably an NHL2 mortar was chosen to match an existing mortar. It does not make sense to increase the strength of the mortar to suit the climate, and to forget about the structure. It would be better to delay the work until spring, or to take additional precautions to protect the work after the NHL2 mortar has been used. Covers, insulation and heat may be required.

HAND MIXING

Q. I have to mix a natural hydraulic lime and sand mortar mix by hand on a small conservation project. How should I do this?

A. Natural hydraulic lime comes as a hydrate in a paper sack and is easily mixed with sand. To avoid varying water contents in mixes and over-wet mixes, a dry sand is best.

First, the lime and the sand are measured carefully by quantity. A plastic bucket is useful for this. A 1:2 mix would be one bucket of NHL to two buckets of sand. Hand mixing is done either by hoe or by shovel.

MIXING BY SHOVEL

To ensure an even distribution of lime throughout the mix, the dry lime and the sand are turned over and back at least three times with a shovel. A trough is made in the centre

Water is added and a hoe is used to mix the mortar

The traditional Irish practice was to use the back of a shovel for mixing lime mortars, not the front. This was much harder to do but gave a better mix and required less water

Use body weight where small quantities of mortar are required that have a low water content

and clean water added. The inside edges of the trough are pushed into the water bit by bit. Add more water, as needed. The amount of water will vary according to how dry the sand is and the purpose of the mortar. A mortar for pointing stone will require a stiffer mix than one that is to be used to build stone. Stiffer mortars require greater physical effort to mix by hand, but it is effort that is required – not additional water.

In Ireland, good hand mixing was always done with the back of the shovel, and not the front (which is easier). Older workers remember when someone who did not work with the back of the shovel was frowned upon and would not find work at mixing mortar. The tradition probably is a carry-over from working with lime putty. This needed to be beaten or compressed into the sand without the addition of any water.

Where a tiny quantity of lime mortar is required for specialist repairs and a fine sand is needed, the lime mortar is sometimes beaten with a hammer in a small hessian sack. This keeps the water content very low, increases workability, and prevents shrinkage on drying out.

Even body weight can be put to good use where small quantities are required – wear Wellington boots and compress or beat the mortar by doing a form of 'lime dancing'. It really works!

It is obvious that additional time should be spent mixing than would be spent in preparing cement mortars.

If the mix is let stand (under cover) for a few hours and then knocked up again, it will become more workable. This is applicable only to the slower-setting lime mortar mixes.

ADDING CEMENT?

Q. The specification for a new classical-style house in stone and brick calls for the use of natural hydraulic lime mortars. There is no mention of cement. I am thinking about adding cement to make sure that the mortar sets.

A. This is a common misconception. It comes from having experience only of high calcium bagged (hydrated) lime, a product normally mixed with cement and sand to make contemporary mortars. Natural hydraulic lime should not be mixed with cement. If it is, the benefits of using natural hydraulic lime will be undone, and problems of increased and unwarranted strength and loss of flexibility will result. Natural hydraulic lime is designed to be used directly with sand and water. Occasionally, and in special circumstances, it is sometimes mixed with high calcium lime – but never with cement.

Mud (yellow clay) well mixed with sand and water using a hand-held electric machine

MUD MORTAR

Q. I am working on a historic structure. The old mortar between the stones is yellow. I think it is mud. There are small white lumps through it. The architect said that if this material has lasted this long, we should replicate it. Samples have been taken for analysis.

A. It is probably a mud mortar. Mud is variously called earth, *daub* or *dóib* (Irish). A subsoil, it is often yellow in colour, but can also be brown or grey-blue (from beneath peat bogs) as well. Used frequently in the past, it makes a very workable mortar, not unlike the wonderful workability of hot lime mixes described earlier (see page 58). Mud was used as a wall building material, a mortar for building stone, a flooring material and a plaster.

Stickiness or tackiness is a key factor in workability. Mortar that will stay on the blade of a trowel turned upside down is a key indicator — mud mortars are excellent in this regard.

To prevent shrinkage, sand often has to be added to the mud. The amount of sand depends on what sand already exists in the mud.

A disadvantage of mud mortar is that in a wet climate, it can get washed out of walls eventually. In the past, with

A nineteenth-century wall built with mud and a small quantity of quicklime. Originally, this would have had a protective lime coating. County Roscommon

thick walls averaging 600 mm plus, there was little need for adhesion between stones — unlike modern walls which average 100 mm. Like weaker lime mortars, mud kept stones apart more than it stuck them together. Quicklime was sometimes added to mud for a number of good reasons:

- To stiffen mud mortars that were too fluid for building.
- To harden mud mortars more quickly so that work could continue without interruption.

When used to build external walls, a lime render was normally applied to prevent loss of the mud mortar between the stones. In the past, mud was cheaper to use than lime mortar.

COBBLESTONES

Q. Cobblestones, small in size and forming decorative patterns as part of an eighteenth-century garden, are laid in a white mortar, set hard. As stonemasons, we are contracted to carry out repairs, in particular to select similar cobbles and lay them to the same pattern in mortar to replace a missing section. What mortar and method of laying should we use?

A. It is very likely that the original mortar was hydraulic or pozzolanic because the mortar needed the ability to set under ground.

Cobble paving at Mayglass, County Wexford

Right: Cobbles newly laid in mud and being grouted in mud, sand and grit. Kilmallock, County Limerick

As far as possible, cobbles are generally laid with their length running into the bedding mortar. Because of the different lengths of cobble, some will have more mortar under them than others.

Check the sand size in the original mortar and replicate. The mortar should be kept stiff.

Two boards, on edge, each side of an area of work with a string line as a guide will ensure accurate levels and grades. Follow the existing pattern – cobbles should be reasonably tight to each other to prevent future movement.

Cobbles in farmyards, larger in size than above, are commonly laid in mud, although sand is an easier medium.

Setts, as seen on the streets of Temple Bar in Dublin, are often referred to as cobbles, which they are not. Setts were cut by hand from whinstone (basalt) to rectangular shapes and laid in courses, transverse to the direction of traffic.

BRICKWORK

INTRODUCTION

Due to the abundance of stone, little brick was used for building in Ireland before the seventeenth century. The craft of the Irish bricklayer didn't develop fully until the eighteenth century when there was a boom in building, particularly in Dublin.

Brickwork remained popular throughout the nineteenth and twentieth centuries. Although associated mainly with the city, it can also be found on many rural buildings, although it is nearly always hidden behind plasters and renders. Brick is also hidden behind stone facings on great country houses and public buildings in cities.

Many older, handmade *clamp* (a type of primitive kiln constructed of the bricks being fired) or *kiln-fired* (a purpose-made permanent kiln) bricks are quite soft and porous. The appropriate mortar to use when relaying or repointing walls built of such bricks should always be softer and more permeable than the brick.

The repair of old brickwork is dominated by the appropriate style of pointing or repointing, and the type of materials used. Dublin bricklayers who specialised in pointing were known as *wiggers*.

Tuck pointing can be seen on many of Dublin's older brick buildings. It is not exclusive to Dublin but the majority of tuck pointed buildings are to be found there. There is some controversy about when tuck pointing began in Dublin, although it was probably in the eighteenth century. There are also questions about whether it is contemporary with the original construction of the brick building or whether it was carried out at a later date as a re-pointing exercise. Terminology also varies but now, with increased research, there is focus on this. Tuck pointing appears to have continued through the nineteenth century and even up to the 1950s. It then disappeared, to be revived in recent years for the conservation of brick buildings.

New tuck pointing, North Great George's Street, Dublin

The remains of gauged brickwork. Dublin

Wigging, also called bastard tuck pointing, was common in Dublin. It shows the projecting white mortar tuck, and red mortar stopping, disguising a wider mortar joint behind. A red colour wash is often applied to the brick face as well

A high level of craft skill is required to carry out this type of pointing and it is difficult to find skilled bricklayers who can carry out tuck pointing correctly.

Tuck pointing consists of a thin projecting bead of white lime mortar which is commonly called a *tuck*, although it is more correctly the ribbon. The tuck averages 3–6 mm in width and projects by 3 mm at most. Except for the tuck, the remaining part of the face of each mortar joint is disguised with a pigmented *stopping* (coloured mortar) to match and repair the brick behind. To obtain a uniform colour, the entire wall was often, but not necessarily, colour washed with an earth pigment of *Red Ochre* or *Venetian Red*. Yellow brick buildings were sometimes colour washed in red as well.

Tuck pointing is a pretence or imitation of something more grand called *gauged brickwork*. In Britain, gauged brickwork was sometimes used for the entire façades of finer brick buildings. Special bricks of consistent quality, with a relatively even and matching colour called *rubbing bricks* or *rubbers*, were used.

Rubbing bricks were purposely made over-sized, then rubbed on a piece of sandstone square on a bed and one face. They were then placed in a cutting box where they were cut to the required length and thickness with a twisted wire saw and then rubbed on the remaining bed. This meant that when they were laid, it was with the minimum amount of lime mortar. Very little gauged brickwork is seen in Ireland.

If finances did not allow for gauged brickwork, or when confronted with hand-made bricks which were much coarser and usually of uneven colour, tuck pointing was introduced to make the joint sizes look smaller and the brick look more refined. Everything other than a white bead of projecting mortar in the brick joints was often colour washed with Venetian Red or Red Ochre. The finished work emulated the far more expensive, refined and accurate gauged brickwork.

Tuck pointing is very much a specialist activity, requiring some tools which are homemade. One of these,

A Frenchman is used for cutting the tuck/ribbon

the *'Frenchman'*, was a kitchen knife especially shaped, bent and sharpened to cut the profile of the tuck. All mortars used contained lime and fine sand, with white marble dust used occasionally instead of sand.

English tuck pointing, which is most common in Britain (there are some examples in the United States and Australia), is different from wigging and does not seem to be common in Ireland.

WIGGING (BASTARD TUCK POINTING) – DUBLIN

Wigging showing loss of coloured stopping each side of the projecting tuck

The following is the method for carrying out wigging. Always work from the top of the building down.
- Rake out the existing mortar joints to a depth which is the width of the joint.
- Brush out all debris.
- Pre-wet the face of the wall with water.
- Then *colour wash* (assuming that there is evidence on the existing building that a colour wash was used previously) the whole brick façade to achieve an even colour throughout, disguising different coloured bricks so that they all look the same. The pigment to use for the wash depends on the required finish and the colour tone of the brick. Venetian Red or Red Ochre are the most common pigments. Other pigments may be added to achieve the desired colour. Pigments need to be fixed, or they will quickly wash off the face of the wall and stain the white tuck joint. *Alum* added to hot water is the traditional fixative. The pigments are added to this. (Colour wash is not a limewash. Limewash is fixed by the carbonation of its lime content; colour wash contains no lime.)
- Pre-wet the raked out joints to prevent the mortar from drying out too fast.
- Insert the lime mortar stopping (un-pigmented), finishing very slightly back from the face of the brick. This is executed with a flat jointer with a width just slightly less than the mortar joint.

- This insert is very lightly double struck using a small trowel.
- Mark/rule the line of the tuck on beds and *perps* (perpendicular or vertical joints) very lightly to create a shallow groove with a depth of 1-2 mm. This groove acts as a guide for the accurate placement of the tuck; it is not intended to be a rebate into which the tuck is inserted.
- Using the same mortar as the stopping, lay on the tuck to the groove using a *brick jointer*. Do this on the same day while the first mortar insert is still fresh, but stiff. This is very important – otherwise the tuck will not combine with the stopping and will fall off. The tuck should project past the face of the brick slightly, no more than 3 mm.
- Using a Frenchman, cut the tuck accurately and uniformly to a consistent width. The width can be 3-6 mm, but must be consistent throughout. Form the tuck on beds first and then vertical joints. The mortar needs to have a fine sand or it will not cut cleanly on the edges. Sand graded from 600 microns (1000 microns = 1 mm) down to dust and of a light colour is about right. The work must be protected from the weather and prevented from drying out too quickly.

Top left: Wigging 1 – Joints are raked out and the brick face colour washed
Top right: Wigging 2 – Plain lime stopping inserted and ruled as a guide for the application of the tuck. The guide is not intended to be a rebate into which the tuck is inserted
Bottom left: Wigging 3 – The tuck is applied while the stopping is still fresh
Bottom right: Wigging 4 – Coloured stopping is applied over the plain stopping each side of the projecting tuck and blended in with the colour wash

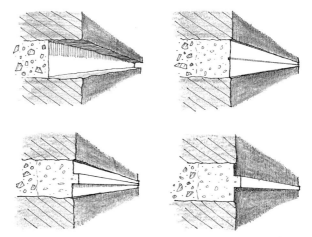

- When the tuck is firm to the touch, apply a paper-thin, pigmented lime mortar stopping on each side of the tuck. This should match the colour wash and be flush to the brick in order to hide the wider, un-pigmented stopping mortar joint. While doing this, repair any broken arrises on the bricks. To achieve the desired colour, this pigmented stopping is composed of lime, fine sand and a variety of pigments such as Venetian Red, Red Ochre, Yellow Ochre and Vegetable Black.

- If necessary, very lightly texture the pigmented stopping to match the brick face by lightly brushing, ensuring no damage occurs to the tuck.

- The pigmented stopping can be touched up with the colour wash in order to achieve a good colour match to the brick.

- Protect from the weather both during the work and for an appropriate period afterwards. Rain can wash pigment down the face of the wall, staining the white tuck.

- Initially, the appearance will be very colourful and sharp. However, the colour wash will fade over time and give a wonderfully soft, weathered look, the remnants of which can still be seen on Georgian brick buildings in Dublin city. No modern materials or finishes can match this delightful effect. Much of the recent tuck pointing carried out is less subtle in quality of execution and finish and does not always complement the original it seeks to recreate.

An alternative to the method described is to insert the plain, un-pigmented mortar stopping in one operation to include sufficient mortar to form the tuck. Mark out the tuck on the face of this stopping and cut away excess mortar each side of the tuck, leaving it projecting proud. A pigmented lime stopping is then used as before to camouflage the wider mortar joint each side of the tuck.

An alternative method of wigging. A full joint is laid on, the tuck marked, and excess mortar removed using the Frenchman. Coloured stopping is now applied each side of the tuck as before

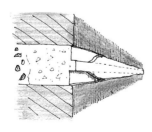

TUCK POINTING – ENGLISH STYLE

English style tuck pointing is quite different from wigging and is therefore shown here in order to clarify. It is mistakenly used in Dublin to repair buildings that were originally wigged.

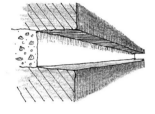

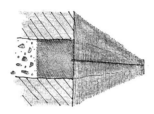

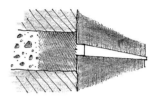

English tuck pointing 1 – Joints are raked out and the brick face colour washed

English tuck pointing 2 – Coloured stopping is inserted and ruled as a guide for the application of the tuck

English tuck pointing 3 – The tuck is applied while the coloured stopping is still fresh

DOUBLE STRUCK POINTING

Double struck pointing, Dublin

Double struck – or more accurately 'ruled top and bottom' – is a flat form of pointing which had its origins in attempts to recreate a neat grid of precise joint profiles like tuck pointing. It was much quicker to execute than tuck pointing and also wider, averaging 6-10 mm in width. It was traditionally finished flush to the brick, but often projects very slightly in modern versions, which is incorrect.

The profile is cut accurately with a sharp blade on each side of the joint to form parallel joints of equal thickness. It is generally applied to brickwork built from reasonably good bricks and controlled joint sizes.

Vertical joints or *perps* are often pointed narrower than the bed joints. This is accomplished using *stopping* (pigmented mortar that matches the brick colour) which camouflages the actual wider perp joint behind. Stopping is also used occasionally to repair brick arrises.

The depth of pointing is usually very shallow, as little as 3 mm.

If executed properly and accurately, double struck can be attractive. When carried out with less accuracy as seen in recent repairs, it is less than effective. Common faults include: excess projection and width; joints not parallel; and the use of wrong materials such as cement. At its worst, it can develop into modern *strap pointing* (see 'Stonemasonry' section, page 65) which has no place on historic brickwork.

DOUBLE STRUCK PROCEDURE

Double struck repointing is carried out as follows:
- Assuming that the mortar joints are recessed/weathered back from the face of the brick, the wall is brushed down and dampened. If not, the joints should be raked out to a depth of twice the thickness of the joint before being dampened.
- If required, the bricks can now be given a colour wash (as described in 'Tuck Pointing', page 100), although with this type of pointing, the natural colour of the brick, either red or yellow, is maintained.

Original lime bedding mortar in 250-year-old garden wall, weathered back from the face of the brick

Left: Double struck 1 – Joints
recessed 3mm by natural
weathering or by raking out

Right: Double struck 2 – Double
struck applied to a 3mm deep joint.
Double struck is applied flush to the
face of the brick. Double struck is
cut parallel in width and is
approximately 8mm wide

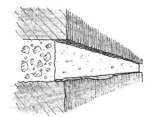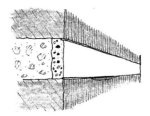

- The mortar for the pointing is of lime and sand. The sand size needs to be fine (600 microns graded down to dust) so that it can be cut with a clean arris to finish.
- The pointing is now applied flush to the brick.
- The pointing is cut parallel using a long straight-edge, the same width as the bed joint of the brick, commonly 8 mm to 10 mm wide.
- Vertical joints are cut using a short straight-edge as a guide. They are sometimes cut slightly narrower than the bed joints. In this case, coloured stopping is sometimes applied to hide the wider joint behind. Only very occasionally is stopping used to repair brick arrises on the horizontal joints.
- The work is protected from the elements and lightly sprayed with water, as required, so the mortar does not dry out too rapidly.

POINTING – FLUSH/BEATEN FINISH

A flush/beaten finish is very slightly recessed from the face of the brickwork by beating with a bristle brush. It is used to replicate original bedding mortar that has lost its original profile and/or *laitance* (lime-rich surface) to the joint, and has slightly weathered back from the face of the brick, exposing the aggregate in the mortar.

The procedure follows a number of steps:

- Assuming that the existing mortar should be removed, this is done to a depth of approximately 25 mm. If it is original, has not deteriorated, and is still reasonably flush, it should be left as found. The fact that an existing mortar is soft is no reason for its removal. If it is sand and cement pointing and is now giving rise to problems, it should be removed without damage to the surrounding brick. As a rule, *disc-cutters* should never be used because of the damage they can cause to brick (widening of joints, over-runs into adjacent bricks, and horizontal wave-like cuts). Only a skilled bricklayer experienced in the use of disc-cutters, and using a small diameter blade with a thin profile, can execute an accurate centre cut to each mortar joint without damaging any bricks. A sharp tungsten-tipped chisel is then used each side of the joint to remove the sand and cement pointing without damaging the brick. Raking out begins at the top of the building and progresses downwards.
- Joints are brushed clean of debris, starting at the top of the building.
- The joints are dampened, ready to receive the pointing.
- The pointing mortar is mixed. In the case of a high calcium lime mortar, lime putty and sand should be mixed well in advance and allowed to sour out in a wet state under cover (page 55) for a number of days at least. This results in higher workability. Where a pozzolanic set is required, a pozzolan is then added to the mortar just prior to using. Alternatively, depending on circumstances, a natural hydraulic lime mortar may be considered. All mortars should be softer and more permeable than the brick. The sand size, colour, shape and grading are all important. A sand in an existing lime mortar may have to be replicated, in which case grain size, colour and shape will be determined. In general, a well-graded sand, with a maximum particle size

one-third the joint size, is desirable – but it may have to be larger if replicating an existing sand. The mortar consistency for pointing work, as opposed to laying, should be quite stiff, yet workable.

■ The pointing mortar is applied using a flat jointer so that it is compressed into position. Vertical joints require the mortar to be sufficiently workable and cohesive so that the bricklayer can lift the mortar off the pointing hawk and onto the flat jointer before application. The mortar is then applied and

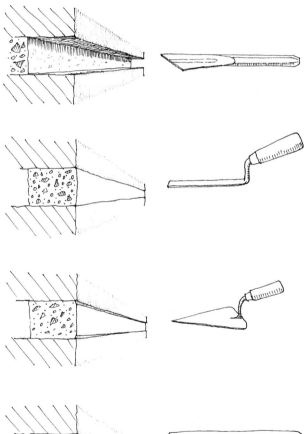

Flush/beaten 1 – Joints raked out using a plugging chisel and hammer

Flush/beaten 2 – Mortar application using a flat jointer

Flush/beaten 3 – Double struck, very slightly before beating, in order to produce a defined brick arris. This procedure is optional and may not be necessary

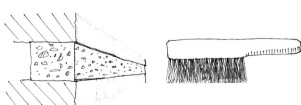

Flush/beaten 4 – Joints beaten with a churn brush

firmed in until flush, or slightly more than flush, to the face of the brickwork. It is important that no indents, holes or cracks are left.

- At this stage, the mortar is left under protection from the elements until quite stiff/hard. It is then beaten with a stiff brush (a churn brush is commonly used). This exposes the particles in the sand, gives a uniform textured finish, slightly re-compresses the joints, and assists in carbonation. Before beating, the joints are sometimes very slightly double struck in order to achieve a more defined brick arris.

- The time between application and beating varies according to the brick, the type of lime used, its water content, and the prevailing weather. If beaten too early, the aggregate will not be exposed and mortar will be removed on the bristles of the brush, thereby staining the brick. If left too late, the beating will not have the desired impact and the joints will have to be scraped with a metal bladed tool; shrinkage cracks will not be closed and the correct standard of finish will not be achieved. The final finish is very slightly recessed, approximately 1 mm or less.

- On completion, the work must be protected. This is often the most neglected part of any pointing process. Lime mortars are slow setting. They require more care than cement-based mortars. Damp hessian and/or plastic sheeting (with a suitable air gap) may be used, depending on the weather conditions, to control the loss of water from the newly applied mortar. Light spraying may also be necessary. If drying takes place too rapidly, the mortar will turn to dust and fail. Protection from frost may also be necessary, depending on the time of year.

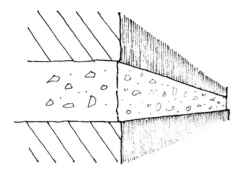

Flush-dragged joint finish

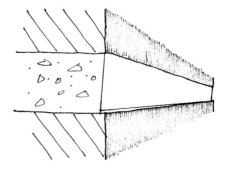

Struck joint finish

FLUSH-DRAGGED JOINT FINISH

This finish is generally created by the dragging action or cut of the trowel edge during the process of laying and trimming the extruding mortar joint. It is not a pointing style, but rather a joint finish.

A full joint with a cut finish in lime mortar is attractive and usually the sign of a skilled bricklayer. This type of finish is not polished or smooth. Instead, it is open and textured, allowing carbonation of lime mortar to take place more completely.

This finish is sometimes found on internal brickwork that was later plastered. When seen externally, it is usually weathered back slightly from the face of the brick. There are similarities to flush/beaten pointing (page 63).

STRUCK JOINT FINISH

This finish is seen on brick walls which are generally plastered or rendered afterwards. In Ireland, it is rarely seen on finished face brickwork. *Struck jointing* is carried out during the process of laying and is generally not executed as a pointing exercise.

To produce this finish, the bricklayer strikes the joint of the bedding mortar during jointing or laying with the trowel, creating a small ledge on the top arris of each lower brick. This ledge forms a key for the first coat of render or plaster.

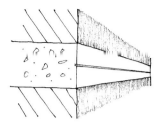

Above: Ruled or penny struck joint finish/pointing

Above right: An early eighteenth-century example of brick and lime mortar. Bishops Palace, Cashel, County Tipperary

Right: Original lime mortar showing a ruled/penny struck joint in Flemish bond. Little deterioration has occurred to either the brick or the lime mortar in nearly 300 years

RULED OR PENNY STRUCK JOINT FINISH/POINTING

This is a mortar joint with a rebated line ruled into the centre of a flush or struck flush joint. *Ruled* or *penny struck* was executed either as a jointing or a pointing process.

INAPPROPRIATE POINTING STYLES AND JOINT FINISHES

A number of inappropriate pointing styles and joint finishes have been used since the beginning of the twentieth century.

Weather Struck and Cut Pointing

Weather struck and cut pointing was common on new work, particularly from just before the mid twentieth century. The example shown is slightly struck at the top and flush

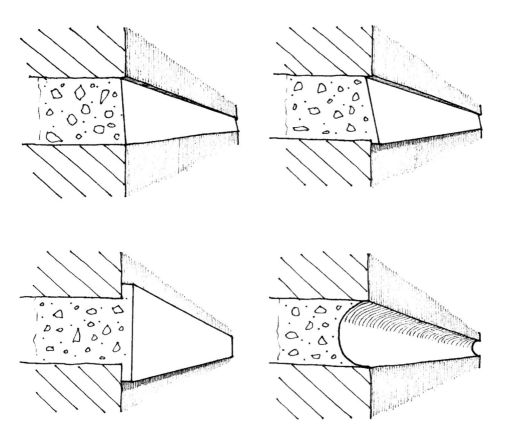

Top left: Weather struck and cut pointing. Inappropriate on old brickwork

Top right: Weather struck and cut pointing. Inappropriate on old brickwork. In this example, the bottom edge of the pointing projects beyond the face of the brick, throwing a shadow, and does not constitute good practice. In some examples, the bottom edge hides a considerable amount of the surface area of the brick

Bottom left: Strap pointing. Inappropriate on old brickwork

Bottom right: Jointed. Inappropriate on old brickwork

with the arris of the brick underneath, which is considered best practice for modern work. The mortar joints are raked out and then pointed using a pointing trowel to finish. Weather struck and cut creates a weathered or sloped finish, the intention being to throw rain off the wall. It is inappropriate on old brickwork.

Strap Pointing

This is a recent innovation in which wide projecting mortar joints are generally of equal width throughout. It is more commonly seen on stonework, where it is also inappropriate.

Jointed Finish

A *jointed finish* on modern brickwork is executed during the laying process using a jointer such as a round steel bar; it gives a

Old clamp fired bricks rebuilt with joints raked out. Inappropriate finish on old brickwork

concave, compressed joint finish. Generally used with cement mortars, it compresses lime mortar and slows carbonation, has no precedent, and is inappropriate on old work.

Raked Joints

This is a modern form of joint finish where the mortar joints are raked out to a uniform depth. It is inappropriate on old brickwork.

JOINTING

Q. I am used to jointing brick with a round steel bar. It gives a nice compressed joint that keeps the rain out. However, the contractor tells me that this is not acceptable on the nineteenth-century house I am pointing. Why?

A. The concave joint created with a round steel bar of 10 mm diameter was first introduced to Ireland about the 1960s. It allowed bricklayers to finish their work as they built and was more economical than having to rake out and then point (weather struck), as had been common practice on new brickwork up to then. Jointing produced a tight, compressed joint in cement mortars.

Although this method is an efficient, economical and technically sound way of finishing modern brickwork, it is inappropriate for an old brick building.

POINTING – WEATHER STRUCK AND CUT

Q. As an apprentice, I was trained to do weather struck and cut pointing in sand and cement. Why is this not acceptable on an old brick building?

A. Weather struck and cut pointing was, and still is, taught to all bricklaying apprentices. It became increasingly popular during the twentieth century in Ireland. It is not appropriate to use this method on older brick buildings as it distorts the visual appearance completely by emphasising the bottom arris of each brick, in shadow, while suppressing the top arris. Impervious sand and cement mixes are used in its execution, preventing evaporation of water from solid brick walls. Unfortunately, weather struck and cut pointing is still commonly used on older brick buildings, irrespective of the period of construction.

Within living memory, newly laid brickwork was raked out at the end of every day. When the brickwork was finished, the building was pointed from top to bottom,

Since the 1950s, the mortar used for pointing tended to have a high cement content, with the erroneous intent of waterproofing the joints. A pointing trowel was used to create a sloped weather struck finish that was considered best for shedding water off a wall. Comparison was wrongly made with slates on a roof.

These impermeable mortar mixes have since created problems. Today, we can see pointing partly missing, having cracked, separated and fallen out from between brick joints, often taking part of the brick with it. Frost damage on the face of bricks is evident too from this type of pointing because of reduced evaporation of water from the wall.

Weather struck and cut pointing had no precedent in eighteenth and nineteenth-century brickwork and therefore should not be used for repointing.

SAND – SIZE

Q. What size of sand should I use for rebuilding old brickwork?

A. Traditionally, sand was sourced locally, with the largest stones removed. Generally, the maximum grain size in sand for a modern lime-based mortar is about one-third the joint size. While this can increase up to half the joint size, anything bigger than that will create problems. The common 10mm joint used in modern brickwork can comfortably take a sand with a maximum particle size of 3 mm, but it is commonly just over 2 mm. Good sand is graded from the maximum particle size, in a series of gradually diminishing grain sizes down to fine dust.

Old handmade clamp and kiln-fired bricks could vary greatly in size and shape. When this occurred, they would be laid with large mortar joints, sometimes 20-25 mm in size, to accommodate this. Coarse sand is often observed in these larger joints. Old well-made bricks of accurate dimension may have joints sizes averaging less than 8 mm.

Whatever sand is used, it is important to have a well-graded range of particle sizes, from the maximum down to fine dust. This controls the amount of lime in the mix and also helps create the varied porous matrix that will allow the lime mortar to flex and breathe.

CHIMNEY REBUILD

Q. I have to rebuild a brick chimney on an old house. The chimney has two square, parged flues. It is in very bad condition and in danger of collapse. The top of the chimney is corbelled. The bricks are soft and damaged by frost. The chimney comes through the roof ridge and is about 1200 mm high above that point. What mortars should I use, and what additional problems should I look out for?

A. These suggestions will assist your work.

- Observe the dimensions and style of the chimney closely, particularly the corbelling style near the top or head. These details should be maintained.

Brick chimney under re-construction using salvaged old clamp fired brick and a natural hydraulic lime mortar. Timber profiles are fixed to the scaffold to give plumb, corbels, course heights and level

Take photographs, measurements, and make a sketch of the bonding arrangement.

- Before starting, jam a hessian sack or similar in the flue to prevent light debris like old mortar falling down the flue. Then carefully strip the chimney down as far as necessary.

- Use a brick of similar dimensions, texture, colour and size, but reasonably well burnt and sound.

- Install lead flashing and a *tray* (a flat lead tray that prevents water penetrating downwards from the chimney to the inside fire-breast) at the point where the chimney comes through the roof. Leave *weep holes* (open vertical joints) above the tray.

- Bond the *withe* or *mid-feather wall* (dividing one flue from another) into the chimney. Don't have straight or running joints at this critical construction point, even if it was done originally.

- Due to the exposed nature of this element of the structure, a pozzolanic or natural hydraulic lime should be used. It should also be selected to suit the bricks (sound and well burnt), the location (severe), and thermal expansion and contraction.

- In the absence of using a *flue-liner* (a pre-formed terracotta pipe), *parging* (plastering) of the flue should be carried out as the work proceeds by

After the mortar joints have been finished by beating with a stiff brush, the work is cleaned down, dampened and protected from the weather

holding a wooden lath as a guide on the top bed of the bricks. The parging is screeded to the edge of the lath using a trowel. Fresh cow dung is often added to the mortar mix at a ratio of approximately 1:4.

FIREBACK – MORTAR

Q. I am working on an old house in which I have to rebuild a large fireback in clay brick. The original mortar looks like a lime mortar, but it has now deteriorated, as have the bricks. What bricks and mortar can I use?

A. Use solid clay bricks rather than the *frogged* (indented on the bed) or *perforated* (holed) type. The longest-lasting bricks for this location are fireclay bricks, which are normally yellow in colour and made from a special *refractory clay.* An ordinary, well-made solid clay brick will last quite a long time.

The following comments describe a number of mortars with varying characteristics that may be suitable for this job. Much depends on how often the fire will be used and the temperatures reached in the firing. Firebacks are often built for easy replacement in the knowledge that they have a limited lifespan. Herringbone pattern brickwork is associated with larger firebacks in some countries.

Portland cement should not be used because of its inability to accommodate thermal movement.

Consider the following:

- A proprietary refractory cement mortar is best for bedding fireclay bricks, but the joint sizes must be very thin, and the bricks need to be made accurately for such fine joints. The bricks are dampened and the mortar bed and vertical joint applied to the brick in the hand before it is laid.
- Fireclay with finely crushed clay brick (firebrick) should work well.
- A good mud mortar with sand (no limestone particles) would allow joints of variable width (the smaller the better) to suit uneven bricks. This will have a limited life, however.

- An NHL1 or NHL2 lime mortar with an appropriate sand might work, but it too will have a limited life.
- A high calcium lime mortar with a brick pozzolan and appropriate sand could be used, but would also have a limited life.

In some areas of the country, firebacks were reconstructed annually using a local mud that was cut in a brick shape from a known earth bank. These were laid dry on top of each other to form the fireback and went hard when burnt. They only lasted about a year. In other instances, mud and cow dung were mixed together to form firebacks.

BRICK NOGGING

Q. I have to rebuild a brick nogging partition. Could you tell me more about this?

A. This dates back to the Middle Ages and is, in essence, a timber-studded partition which is in-filled with clay brick, built in lime mortar (high calcium), and plastered in lime and sand. The partitions are normally 4½ inches (112 mm) thick.

The bricks are laid either horizontally in a stretcher bond or in a herringbone pattern with their faces flush with the timber framing. Brick noggings (in-fill of brick built in lime mortar) help make the partition more robust and improve its fire and rodent resistance, as well as the thermal and acoustic performance.

The joints of the brickwork were frequently struck to act as a key to hold the plaster that was applied to both faces of the partition. Brick bedding mortars and plasters for this type of work were all lime based. Hard turf (peat) was sometimes used instead of brick. Brick noggings are known to survive in a few buildings dating from the seventeenth to the nineteenth centuries.

SPALLED BRICK

Q. A brick building I have to repair has many spalled (face surface of brick partly lost) brick faces. It dates to the early 1950s and is pointed with a weather struck and cut joint in sand and cement. I have to repoint it. The bricks are a light red/orange colour and fairly soft. The brick outer leaf is 225 mm thick. Why has the spalling occurred, and what mortar should I use for the repointing?

A. When this building was built, the procedure was to rake the mortar joints out at the end of every working day. When the brickwork was finished, the building was then pointed from the top down, often by the bricklayers who built it but sometimes by specialist pointers or wiggers.

Although the bedding mortars were lime rich and relatively soft, often with little or no cement, the pointing mortar generally used had a high cement content, sometimes with a little lime to make it more workable. The intent was to achieve an attractive finish and, erroneously, to waterproof the joints and the wall.

Problems occurred particularly with softer bricks. Bricks pointed with a cement mortar do not dry out as quickly as do bricks pointed with lime. Soft, wet bricks are vulnerable to spalling from frost action, and from the wetting and drying cycles when salts migrate through the bricks.

Removing cement mortar is difficult without damaging the arrises of the brick, particularly when the bricks are soft – in fact, it may not be possible to do this. There is a specialised technique using a very small angle grinder with a thin blade which cuts the centre of the joint. A small, sharp chisel is then used to remove the pointing each side of the cut. This works well only when carried out by a skilled bricklayer who specialises in the technique, but is very destructive otherwise because of overruns and widening of joints. Removal using very sharp masons' chisels is very effective, but only if a highly skilled and experienced brick-layer is employed.

The replacement mortar for the new pointing needs to be more permeable than the soft brick. A feebly hydraulic lime mix could be considered.

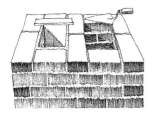

Above: Parging of the flue with lime, sand and cow dung

Right: An iconic symbol of Dublin: a windmill built in brick and lime. The former Roe's Distillery, Thomas Street, Dublin

Far right: Detail photo of the windmill showing the compatibility and success of a soft brick laid in an appropriate mortar

CHIMNEY PARGING – LIME AND COW DUNG

Q. What mixes were used for parging (plastering) the insides of brick and stone chimney flues which incorporate cow dung?

A. Many traditional lime mortar mixes are approximately 1 part lime to 2 or 3 parts of sand, although this can vary widely. The amount of cow dung added to the mix also varies, but it could be up to 25% of the lime mortar. Hair was sometimes added to the mortar to reinforce it and to prevent cracking.

Cow dung not only aids adhesion; it also helps to resist the sulphates created by soot and water from penetrating the wall and causing damage to mortars, bricks and plaster and producing brown nicotine-like staining. Parging also prevents smoke and fire from getting through any openings or cracks in the mortar joints of the brickwork. Over time, parging is usually lost and sulphates are seen to penetrate the fire breast. These show up as a dark, sometimes greasy stain on the chimney breast.

If the plaster on the chimney breast has to be replaced, the addition of cow dung to the lime mortar helps to alleviate the problem.

PLASTERWORK

INTRODUCTION

Ireland's great period of internal decorative plasterwork took place from the seventeenth to the nineteenth centuries, with some of the finest examples in Europe executed during the eighteenth century. Irish plasterers quickly adopted the skills of immigrant *stuccodores* (plasterers) to produce their own unique style. Their work can be seen in all its magnificence in Dublin's eighteenth-century interiors and in large country houses. Many fine examples also exist in the smaller Georgian and Victorian buildings, generally executed by unknown artisans.

New render finish with a finely ruled joint to imitate ashlar on the former St Mary's Church, Mary St, Dublin

Left and right: Decorative
plasterwork, Russborough,
County Wicklow

Throughout this text, the term *plastering* refers to internal work, while *rendering* refers to external work.

On close examination of old plasters, it can be seen that although the finish coat is lime putty (sometimes pure lime putty without sand – generally not recommended), the backing coats are often hot lime mortars (see 'Stonemasonry', page 58). In this situation, it was advantageous to allow the hot lime mortar to sour out over a prolonged period of time before using it, thereby ensuring that un-slaked particles of quicklime would not spoil the work later. As a further complication, lime putty was sometimes run into hot lime mortars to produce even fatter and more workable mixes when required.

In the first half of the twentieth century, high calcium lime as a lime putty, hot lime mortar, or crude hydrate were replaced with a more refined industrial processed hydrate that came in a sack. Lime-only mortars were then superseded by lime hydrate gauged with cement or gypsum. Cement, gypsum and plasterboard eventually began to take over, with the eventual near elimination of lime from the mix altogether. Applying these modern mixes to older buildings created problems. In order to repair or to replicate old plasterwork, it is necessary to use materials and techniques similar to those used originally. It cannot be stated strongly enough that a plasterer who is trained, skilled and experienced in the conservation of plasterwork is essential when it comes to the repair of historic plasterwork.

LIME PUTTY

Lime putty has various names within the plastering craft – plasterer's putty, run lime, air lime, and of course fat lime and rich lime. The present lime putty manufactured in Ireland is classified as a high calcium lime (CL90). Being non-hydraulic, it will not set in continually damp conditions or where access to carbon dioxide is restricted.

Lime putty was very much at the centre of the traditional plasterer's craft, and great emphasis was placed on its preparation, application and after-care. Generally the older the lime putty was before being used, the better the mortar was. Allowing it time to stand in a putty form before use eliminated small quicklime particles that could cause subsequent damage to finished plasterwork by expanding and causing popping.

How to Make Lime Putty
The process of making lime putty is known as *slaking* – when quicklime and water are mixed together to produce lime putty.

A large, shallow wooden box or a large metal bath was placed on the ground and partially filled with clean, cold water. Quicklime was then added, creating an immediate reaction. The lime and water quickly reached boiling point, often within a minute. This highly dangerous activity requires personal protective equipment and training.

A hoe or rake was used to agitate and break down the quicklime in the water. While still hot, the mixture (now the consistency of milk) was run through a sluice gate in the wooden box into a lime pit below ground level. The sluice incorporated a sieve which prevented larger lumps of quicklime and unburnt pieces of limestone from getting into the pit. If no sieve was used on the sluice gate, heavier pieces of unburnt limestone and incompletely slaked pieces of quicklime would sink to the bottom of the pit.

Above: Steam rising at a lime-slaking demonstration at the Building Limes Forum gathering in Cardiff, Wales, 2006

Right: A slaking box and pit being restored for re-use at a disused tannery. The slaking box is in the foreground, with a small opening where a sieve and sluice gate will be installed. In the background is the pit where the lime putty will be stored. Ballitore, County Kildare

When the lime settled, if there was excess water on top of the lime putty, it was bucketed or siphoned off, leaving just a thin layer of water to cover the lime. If the pit was unlined and the water table in the ground was low, the excess water would simply drain away. The lime putty in the pit was left undisturbed for three months or more while it thickened to the consistency of yogurt or cream cheese.

Leaving the putty in the pit for so long reduced the chances of having pieces of quicklime, known as *inclusions*, in the final mortar. This is something the plasterer was most concerned about, particularly on the finish coat, because it could make the plaster pop or blow from expansion.

Lime putty formed by slaking.
Building Limes Forum, Cardiff,
Wales, 2006

On leaving the pit, lime putty performs better in a stiff rather than a liquid state. Excessive water leads to weaker mortar mixes and increases shrinkage.

Re-Hydrated Lime

High calcium lime as a hydrate is sold in a paper sack and is widely available throughout the world. There are advocates of its ability to be re-hydrated to produce lime putty. It is generally considered a material of lesser quality than traditionally produced lime putty. The principal reason for this is that, while in the sack, it has access to air and therefore begins to carbonate before being used. In the second half of the twentieth century when lime putty was no longer available as an everyday product, fresh lime hydrate run to a putty was used in the conservation of plasterwork for fine grout mixes and *in situ* cornice work.

For the most part, experienced practitioners believe that lime putty, slaked traditionally from quicklime, is better. The lime revival of the last twenty years or so now means that lime putty is readily available from specialist suppliers.

SPLIT LATH – PARTITIONS AND CEILINGS

Q. What is the typical procedure used in three-coat lime plasterwork on split laths?

A. In the past, the vast majority of internal work – particularly on lathed ceilings – was carried out with high

Lime plastering to a split lath partition showing first (scratch), second (float) and third (finish) coats

calcium lime putty plasters. The use of natural hydraulic limes on ceilings can result in cracking due to early strength. The gradual deflection of the ceiling that takes place from the weight of the plaster can also result in cracking. The weak and soft high calcium lime plasters are ideally suited for this slight flexing. It should be noted that much of the vernacular work was finished with a paper-thin coat of neat lime putty. This was trowelled on, then brushed over with a stock brush to close the cracks, and sometimes limewashed. This method is not generally advocated because of the likelihood of cracking and failing.

Here is a brief outline of three-coat work on split laths.

■ Laths are fixed in horizontal lines, with gaps between each line of approximately 8-10 mm. It is important that the gap is wide enough to hold a reasonable quantity of lime plaster so that it will not close when the laths absorb water from the plaster. The laths are called split laths because they are split by hand along the grain of the wood. They are strong, flexible, twisted, of uneven thickness, and sometimes have coarse ridges from the exposed grain, all of which makes them more suitable for lime plaster than the more uniform sawn laths. The laths are butt jointed and fixed in bays, then bonded like brickwork to avoid long runs of joints. Ceiling joists should be at no more than 350 mm centre-to-centre. If the ceiling joists are 75 mm or more in thickness, counter laths are nailed along the bottom of the joists to allow for a better key at this point. The laths are wetted before any work begins.

■ The *scratch* or first coat, also known as the pricking up coat, is applied about 10-12 mm thick. It is very important that this coat is well haired. The scratch coat should not be overly wet, but must be plastic enough to pass through the laths. The coat is applied diagonally across the laths, starting at the top in the case of a partition. It is then allowed to tighten up for an hour or two before scratching

Split lath partition ready for the first
or scratch coat of lime plastering

with a lath scratcher. It is scratched diagonally
both ways to create diamond patterns and
accentuate hardening. It is important that this coat
is well scratched around windows and doorways
where there will be vibration. Hardening is
dependent on the weather and can take from two
to three days, longer in winter. A rough field test
that indicates when the coat is sufficiently hard to
receive the next coat is when there is little or no
impression left from pressing with the knuckle of
the forefinger. The work should not be allowed to
dry out too rapidly – a fine mist spray may be
necessary on occasion.

■ The *float* or second coat is about 8-10 mm thick
and also contains hair. The scratch coat will
probably need to be wetted before the float coat is
applied. Once the float coat has been straightened
up, it is then scoured with a wooden float to
condense and counter the effects of shrinkage and
cracking. The surface of the float coat is then
scratched very lightly with a *devil float* – a wooden
float with small nails projecting about 1 mm.

■ The *finish* or third coat, also called *setting*, is
applied to the float coat while it is still damp but
firm; if the float coat is dry, it will need to be
wetted. Lime putty and sand are used for the finish
coat. The sand needs to be fine (about 1 mm
maximum particle graded down to dust). The lime
putty and sand need to be mixed at least 24 hours
before the work commences. The finish coat is
applied in two coats, with the second coat applied
over the first while it is still moist. The first coat is
applied with a skimming float in order to leave an
open surface to receive the second coat. Once the
coats are laid on, the whole surface is well scoured
with a wooden float. This is very important
because it reduces the possibility of *crazing* (small
cracks) and cracking. If the finish coat is drying
too fast, it will have to be sprayed lightly with

clean water. This will also aid the formation of an even texture on the surface. A degree of fat or laitance (rich lime content with little or no sand) will come to the surface. This in turn will be *trowelled down* when the final finish (*trowelling up*) is being carried out with the plasterer's steel trowel.

THREE-COAT LIME PLASTERWORK TO STONE

Q. What is the typical procedure for the internal plastering of a stone wall in three coats to achieve a flat, professional finish?

A. This should help you achieve a professional finish.

- The wall should be clean and free of paint, grease, oil, mould, plant growth etc.
- Open and recessed joints should first be pointed with lime mortar. Large joints should have pinning stones inserted flush to the face of the wall.
- Hollow spots should be daubed out using a lime mortar with hair. Coats should not exceed 10-12mm in thickness. The deeper sections should be daubed out and flat stones or broken clay tiles pressed into the mortar.
- It is better if the faces of the stone are slightly rough rather than smooth, thereby acting as a key for the first plaster coat. The wall should be thoroughly damped with water before proceeding to the next stage.
- A scratch coat or first coat of lime plaster with hair is laid onto the wall face. Either a lime putty and sand mix or one of the weaker natural hydraulic limes is used. This coat should not exceed 12 mm in thickness. It is screeded roughly to eliminate most of the undulations. The edge of the plasterer's steel trowel is then used to leave a reasonably flat but coarse, open surface. It is then

Lime plastering at Parke's Castle,
County Leitrim

Right: The remains of the original
medieval plaster left *in situ*,
surrounded by new lime
plasterwork, at Muckross Abbey,
County Kerry

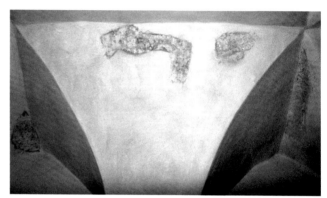

scratched to provide a key for the next coat and to
induce carbonation. This coat is prevented from
drying out too quickly by using a light mist spray
of water as required.

■ When it is difficult to make an impression on the
scratch coat with the knuckle of your forefinger, the
float coat or second coat is laid on over the pre-
dampened scratch coat. For highly accurate work,
75 mm wide mortar screeds are first laid on vertically
with the trowel over the scratch coat. These are
plumbed and laid in range with one another. The
float coat is laid on between the screeds with lime
plaster and animal hair to a thickness of 8-10 mm
and screeded with a wooden *darby*.

- The float coat is well scoured with a wooden float prior to lightly scratching with a *devil float.* If natural hydraulic lime was used in the float coat, a glaze or tight skin may have formed on the surface; this will have to be removed before applying the finish coat. It can be washed off by scrubbing lightly with a churn brush and clean water. Precautions are necessary to prevent this coat from drying out too quickly.
- The finish coat or third coat is usually lime putty and sand laid on in two coats as described in 'Split Lath – Partitions and Ceilings', page 124.

CORNERS AND ARRISES

Q. How can I avoid damage to corners and arrises when using lime putty plaster? In modern work, I use angle beads which give a very sharp edge but somehow this does not seem appropriate in this case – a small, single-storey thatched cottage built in stone.

A. The corners/arrises can be rounded to make them less liable to damage. The walls on either side of the corner are first plastered, finishing the corner to a straight edge. Then a timber template with the desired curve is run by hand up and down the corner, removing excess material until the desired angle is formed. The slight unevenness of this method will leave a vernacular feel. If a truer angle is required, straight edges are fixed each side of the corner and a template is run against these.

Internal plasterwork with rounded corners and arrises

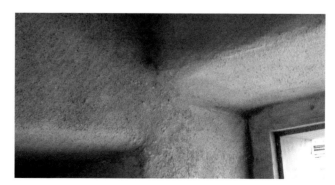

Wooden bead to corner of wall prevents damage to lime plasterwork. Early eighteenth century, Dublin

Sometimes, a permanent wooden bead is found internally in old work at corners, or perhaps as a stop bead beside a frame, a forerunner of the modern metal angle bead. Check for any existing evidence of such details found on site. A gauged lime plaster (see Gauged Lime Plasterwork, page 132) will provide extra strength at corners and arrises.

HAIR

Q. Red and orange ox hair is evident in the internal plasterwork of the nineteenth-century cottage I am working on. Can I still get ox hair nowadays? How much should I use?

A. It is not possible to obtain ox (bullock's) hair in Ireland. As cattle are no longer wintered outside, their coats are not as thick as they once were. Horsehair is once again available in Ireland, and yak and goat hair are imported by specialist providers from China and elsewhere.

Allow about 5 kg of hair per cubic metre of mortar. A sufficient quantity of hair is critical, particularly on lathed work — especially ceilings where it acts as reinforcement to prevent cracking and failure. Hair is also used in plasterwork on walls of stone and brick to prevent shrinkage cracks. Although mostly found in backing coats, it can also be seen in plain and decorative finish coats. Mostly used internally, hair can sometimes be found in some external renders such as at the seventeenth-century Mallow Castle, County Cork.

Goat hair, left and centre; yak hair to the right

The hair should be added to the mix near the time of application. Otherwise, it will degrade while stored in a wet, highly alkaline environment. The hair must be introduced carefully by teasing out the strands loosely rather than introducing it in large clumps. Hair length should range from 30-100 mm in length. When hemp is added to lime plasters, it acts as an alternative to using hair.

LIME PUTTY – SETTING PROBLEM

Q. We used lime putty and sand to plaster the interior of a church. Three months on, the plaster still has not set in certain areas and remains wet. What is the problem?

A. The following are possible causes:

- The lime plaster was laid on too thickly in a single coat. This slowed down carbonation.
- The lime putty plaster was laid on a damp wall that has remained excessively damp. Continual dampness from penetrating or rising damp will prevent carbonation ever occurring.
- The work was protected with plastic sheeting without access to air.
- The surfaces of the backing coats were excessively trowelled rather than left open to induce moisture evaporation and carbonation.
- Carbonation is slow during winter because of low temperatures.
- The wrong mix of lime putty and sand was used.
- The background was flooded with water rather than just sufficiently dampened to ensure dust is removed and suction controlled. The amount of water required to dampen down varies greatly between types of stone and brick.

A pozzolanic or weaker natural hydraulic lime should possibly have been used in the backing coats.

Multiple thin coats are best. Each coat should be applied when the coat beneath it has nearly hardened. Coats should not exceed 10-12 mm each.

Rather than undo what has already been done, it looks like you will have to wait. You could assist the setting process by carefully providing low-level heat in the building. Be careful not to cause rapid drying. Spraying with a fine mist of water occasionally may be necessary, as well as closing any shrinkage cracks with a wooden float as they occur.

If the wall is continually damp in places, there may be another problem – perhaps leaking gutters or faulty downpipes are causing the dampness. The removal of an external render in an exposed setting could also result in excessive dampness.

GAUGED LIME PLASTERWORK

Q. What is gauged lime putty plaster?

A. Plaster of Paris is added to lime putty and sand mixes to produce a mixture called *gauged stuff.* Plaster of Paris/casting plaster is added to create a faster set. The amount of plaster of Paris added to lime putty and sand mixes varies – from equal parts in the case of running *in situ* cornices to one part plaster of Paris to three parts lime putty and sand for setting coats. These mixes will encounter problems if used in damp or moist areas. Once the plaster starts to set, the mix should not be re-worked as this will kill the plaster set.

In the broader sense, the term gauged in plasterwork can mean the addition of many kinds of setting agents such as natural hydraulic lime, a pozzolan, and cement for modern work.

RUNNING A CORNICE *IN SITU*

*Q. What is the procedure for running an **in situ** cornice in lime?*

A. A cornice is a decorative element situated at the junction of the wall and the ceiling. Cornices vary from relatively simple to very elaborate. They are generally classified into three types.

Small cornices – run with no coring out mix (building out the cornice in multiple coats).

Medium cornices – require a coring out mix.

Large cornices – coring out would create too much weight and instead, laths or expanded metal are fixed to brackets.

In Victorian times, lime putty and sand mixes used for coring out cornices were gauged with gypsum plaster. Gypsum plaster was useful because it counteracted the natural tendency of lime and sand mixes to shrink. Care should be taken when repairing old cornices because the addition of gypsum may not be appropriate where it was not used before. Many cornices were created with just lime and sand mixes. The analysis of historic plasterwork by a laboratory experienced in this field, combined with the on-site skills of an experienced plasterer in the repair of historic plasterwork, are crucial to the outcome.

The following is a brief outline for running a plain *in situ* cornice:

- If an existing cornice is to be repaired or replicated, an accurate copy must be made of its profile. A profile gauge is sometimes used but it is not as accurate as making a saw cut in the existing cornice, inserting a 22 gauge zinc sheet, and scribing an accurate profile of the cornice on the zinc. The zinc is then cut, filed to shape and checked against the original. The saw cut is then repaired. Where a saw cut may not be acceptable practice, a rubber mould may be taken which is then used to cast a short length of cornice. This is then cut with a saw and the profile traced onto sheet zinc. For a new cornice, the profile of the desired cornice is either drawn directly on the zinc or drawn on paper first, then transferred to the zinc sheet using carbon paper. The zinc is then cut and filed to shape.
- Both the wall and the ceiling are brought to a straight and plane surface in two coats (scratch coat and float coat). The finish coat on the wall and ceiling is carried out after the cornice is complete.

Brackets fixed to receive a large
curved cornice to be run *in situ*

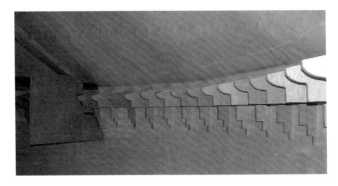

Lime cornice being run *in situ* using
a centre pole, trammel and horse

- The zinc template for the cornice is fixed to a
  timber board which is cut 3 mm short of the zinc
  profile. This ensures that the timber does not swell
  and interfere with the running of the cornice. The
  zinc template fixed to the board is now fixed at
  right angles to another board and braced to hold it
  in position. The whole piece is called a *horse*.

- Large cornices cored out solid would be too heavy
  and would therefore sag. To reduce their weight,
  they are constructed using timber brackets (fixed
  to ceiling joists or wall) to which split laths are
  fixed. The brackets are similar to the profile of the
  zinc template but reduced in size and often of a
  more simple design. The split laths are nailed to
  the brackets with gaps of 10 mm between.

- The horse is run against guides fixed to both the
  wall and the ceiling. The guide fixed to the wall is a
  timber lath called a *running rule* or *slipper*, measuring
  approximately 50 mm x 20 mm.

- The guide fixed to the ceiling is a straightened screed gauged with gypsum to guide the nib of the horse.
- The work is built up in coats. Each of these coats is run using an individual timber template which is cut to shape and fixed to the horse.
- The finish coat is run using the zinc template fixed to the horse. This coat is approximately 3 mm thick and comprised of a lime putty and gypsum plaster. No sand is added to the mix to avoid scoring the surface finish. Typical Victorian mixes were approximately equal parts of lime putty and plaster of Paris. The lime putty was formed into a ring, water was added, and then the plaster of Paris was added and mixed in. Earlier cornices may have no plaster of Paris added, so care should be taken when replicating mixes.

A lime cornice being run *in situ*

A plasterer's small tool being used to form the internal corner of the cornice

Repairing a damaged cornice

Finished curved cornice

Internal corners and areas where the horse cannot be run
are carried out by hand using joint rules and plasterer's
small tools.

WET DASH

*Q. I have been asked to carry out an external wet dash on
an old two-storey stone house using a lime mix. How do I go
about this?*

A. This type of render is called *wet dash* in Ireland, *harling* in
Scotland and *roughcast* in England.

- In the past, the mortar used to wet dash was often
  the same as that used as a bedding mortar to build
  the stone.
- Whether a natural hydraulic lime, high calcium lime
  or a pozzolanic lime mortar is used depends on
  many factors – the need to match an existing render

for historic reasons, permeability, the nature and quality of the substrate, location of the building etc.

- Any pointing and pinning of joints in the wall should be dealt with first. Use a mortar that matches the existing bedding mortar between the stones.

- Hollow spots on the wall face are *daubed out* (filled) using lime mortar and pieces of stone or clay tile. If the wall is flushed out too much, the slightly undulating character, created by individual stones, may be lost.

- A coat of wet dash is then applied to the wall using a harling trowel. If another coat is to follow later, this first coat is flattened with a wooden float by pressing lightly. When firm to the touch, another coat of wet dash is applied. Two or three coats are normal. Old lime dashes vary in thickness, from very thin to over 25 mm. On examination, it may be found that only a single coat was applied. Sometimes multiple coats were applied over previous coats while these were still green (pre-set); this then looks like a single coat. Many of these old wet dash coats were a hot lime mix (see Stonemasonry, page 58) applied either hot or cold and having a high lime-to-sand ratio in the mix. If applied hot, it would have been a very dangerous activity.

- Sand/aggregate for wet dashing should be well graded, from the largest particle down to fine dust. Sometimes, a larger-sized aggregate is also added to create a coarser and bolder texture. This texture is likened to an oatmeal porridge. Any evidence of a previously used sand/aggregate should always be followed for size, colour and shape. A matching sand/aggregate may be available locally. Sharp sand/aggregate is the norm for all types of renders today, although weathered or rounded sands and aggregates were used in the past. They look better and throw off water more efficiently than the sharper particle shapes.

- Protect from the weather at all stages, and be aware that lime mixes set more slowly than cement
- As an integral part of the work, limewashing should be applied in about five thin coats when the wet dash has set. Older wet dashes which have been limewashed have a beautifully soft, rounded appearance. This is due to the build-up of limewash coats over the years, as well as the action of rain and frost softening the sharper elements on the surface. It is difficult to replicate this appearance on new work.

In modern work, the walls (concrete block) are usually *scudded (spatter-dashed)* to form a key. This is followed by a *laid-on* coat (*scratch coat*), and possibly by a second laid-on coat (*float coat*). Finally, a *thrown-on* coat (*wet dash*) is applied.

Modern wet dashes often use a very large aggregate compared to more traditional wet dashes. In modern work, a waterproofing liquid is sometimes added to cement mixes. This evens out any suction and prevents the outline of the concrete blocks showing through when the wall is wet. Waterproof liquids should never be added to lime mixes because it prevents them setting; it also prevents evaporation occurring in solid stone walls, which leads to dampness.

New wet dash on left applied to replicate existing. Kilmallock, County Limerick

Lime wet dash on stone, later limewashed. The restoration of this building in the late 1980s using lime slaked on site is one of the earliest examples of the lime revival that began in that period. King House, Boyle, County Roscommon

Ruled ashlar external render imitating stone

RULED ASHLAR STUCCO

Q. Ruled ashlar stucco has been specified for the front of a three-storey rubble stone building on a street in a rural town. When the original nineteenth-century lime stucco was wrongly removed to show the underlying stonework, this resulted in a loss of character to both the building and the street. The building itself now suffers damp from water ingress through the wall face and drafts around windows where reveal render was removed. The local authority has ordered the stucco to be replaced. Our plastering firm has to carry out the work, ensuring that it matches the original or that it is compliant with the general standard and style seen elsewhere on similar buildings on the street. We must use lime only mixes, something we are not conversant with, and also have little experience of ruled ashlar stucco. Can you help?

A. Ruled ashlar stucco is a smooth external render that imitates finely cut ashlar stone. It was a common style of render generally laid over rubble stone in the nineteenth and early twentieth centuries. It can be seen not only on the small vernacular but also large public buildings. Ruled ashlar stucco can be plain and simple, or combined with projecting decorative stucco details such as quoins, plinth, and run *in situ* moulds at window, door, string and cornice. At other times, the decorative details are in stone, with plain ruled ashlar stucco between.

Before you commence:

- Check for remaining evidence on the face of the building regarding stucco type, thickness, number of coats, course heights, vertical bonding pattern, decorative details at windows, doors, plinth, eaves and corners.
- Check old photos for any details so that these can be replicated.
- Check adjacent buildings where similar details may still exist. It is not unusual that the same plasterer or family of plasterers worked in the same town over a prolonged period and developed their own style.

As ruled ashlar stucco imitates ashlar stonework, certain rules apply:

- Horizontal joints occur at key points – at the top and bottom beds of stone or stucco quoins, cills, window and door heads, springing points and crowns of arches. Horizontal coursing is equidistant, parallel and level throughout the full width of the building. Horizontal stucco joints in adjacent buildings may align with each other and should be continued through your replacement stucco. Projecting quoins in stucco at either end of an intermediate terraced building can usefully break the line of sight and disguise the fact that the plain stucco between them cannot be aligned with the stucco on the neighbouring buildings.
- Vertical joints in ruled ashlar stucco comply to common ashlar bonding rules in stone, such as one-on-two and two-on-one with no running joints.

The vertical joints on alternate courses are often plumb with one another throughout the height of the building, but this can be broken deliberately to reduce the resulting severity imposed on an otherwise less than perfect façade. Projecting full and half quoins in stucco or stone are sometimes shown at wall ends. Arch voussoirs radiate to their true striking points, and the voussoirs themselves are of the correct height in relation to the span.

- The stucco wall face is in range, without hollows or bumps.
- The jointing is narrow, with sharp arrises, straight, and of constant width and depth.
- The face of the stucco is of even texture throughout. Day-work joints are matched to horizontal and vertical ashlar jointing.

It is obvious from the above that the work needs to be planned accurately before the work commences. The wall face and its openings need to be measured horizontally, vertically and diagonally. A drawing is then produced so that the ashlar jointing pattern is predetermined and can be marked on the building.

Before starting, it is important that the substrate is structurally secure, with no loose stones, particularly at openings and parapets. If not, these sections should be dismantled and rebuilt using lime mortar. It may also be necessary to point and pin larger joints with small stones; if so, lime should be used.

The lime mortar mixes can be determined by laboratory analysis from samples of the original. Generally, either natural hydraulic lime or pozzolanic lime mortar mixes are used. It is normal practice that each coat laid on is weaker than the previous.

- The work commences by fixing a vertical timber screed at either end of the building. If the building is part of a terrace and has stuccoed buildings attached on either side, then only intermediate vertical screeds carried out in lime

Ruled ashlar render showing three
coats and vertical mortar screeds

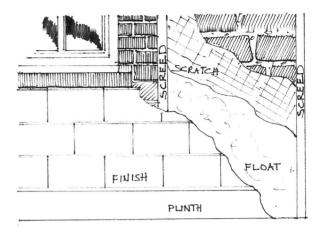

stucco are required. The end screeds are plumbed
and fixed to project past any irregularities on the
wall face. Not all existing buildings are plumb, and
the importance of the screeds is that they are
straight and 'boned in' one with the other rather
than being perfectly plumb. The aim is to achieve a
flat plane surface without excess thickness.

■ A horizontal string line is drawn between the two
vertical end timbers. Mortar spots (in the past,
nails were sometimes used) are established at
approximately 1.5 m horizontal centres between
these timbers. Then a series of vertical mortar
screeds, approximately 75 mm wide, are laid onto
the wall true to the mortar spots.

■ Between these screeds, two coats of lime render
are laid on to a dampened background. The
second coat is flush with the screeds. Hollow spots
may require more than two coats, with excessive
hollow areas needing to be built out using
terracotta tile, broken brick or flat pieces of stone.
All coats are finished with a wooden float and
scratched to form a key for the following coat.
Stucco generally requires three coats. Each coat
laid on is of a weaker mix than the previous.

■ The finish coat is laid on over the screeds to a
thickness of approximately 6 mm. A fine sand (no

more than 2 mm maximum particle size) will facilitate the jointing that is to follow. A horizontal string line is pulled across the face of the building to the course heights marked on the end screeds. The top surface of a long timber straight edge is held to this line, and a brick jointer 3 mm wide is run along the straight edge to form a square recessed joint of just less than 3 mm.

- Vertical joints are marked plumb with the brick jointer against a short timber straight edge. The straight edge is held against a plumb or snapped chalk line. Vertical joints are worked neatly into the horizontal joints. Sometimes, vertical joints are slightly reduced in width and depth to diminish their impact and therefore place more emphasis on the horizontal joints.
- Before the final stucco coat has set, the joints are brushed lightly to remove any rough edges and pieces of mortar that will set hard and make the work look unfinished.
- The work is protected at all stages from excess rain, sun, wind and frost. A light spray of water may be necessary at times, depending on the weather.
- Although many stucco renders can be seen which are painted with modern plastic and enamel paints, it is best to use an exterior grade, lime-friendly breathable paint.

MUD RENDER COAT

Q. The first coat of render on the external wall of an old house in County Kildare appears to be mud. Small particles of coal also seem to be mixed through the mud. Why was this used in the past?

A. Mud was used as a plaster because it was free and was known for being practically waterproof. The term 'mud' is used here, but 'clay' is commonly used too.

The small particles of what looks like coal probably are bits of coal. It is known that mud and anthracite coal were

Right: Culm crusher, County Carlow.
Stones like these were sometimes
used to mix lime and sand

Opposite: Recently restored and
redecorated decorative lime plaster
ceiling in the entrance hall of
Russborough, County Wicklow

mixed together in the Carlow/Kilkenny area for domestic fires. A large stone wheel called a *culm crusher*, fixed to an axle, was pulled by a horse, moving in a circle, to crush the coal and temper the clay. An interesting book on the subject is *Dancing the Culm* by Michael J. Conry, Chapelstown Press, Carlow, 2001.

The culm crusher works on a principle similar to a mortar mill. It was sometimes used to mix quicklime and sand too. The coal in your render is probably there as a consequence of this way of working – tempered mud was needed and the available mud had coal in it.

The remains of charcoal and coal are also evident in lime plasters and renders as a natural consequence of burning lime using wood or coal.

LIMEWASHING

INTRODUCTION

In times gone by, limewashing was applied to stone and mud buildings throughout Ireland. Its use survives within living memory – buying lime at the country fair, adding the lime to the water, applying the limewash by brush. Today, it is still used on farms as an antiseptic in buildings that house animals, as well as by occasional old-timers on their houses as part of a rural tradition unbroken for centuries. The iconic thatched cottage seen on postcards for the last hundred years is invariably limewashed, usually brilliant white but sometimes with a pigment added. Limewash can also be seen on the bigger houses where it was applied to external renders, as well as in basements and outbuildings. Besides looking attractive, the advantage of limewash is that it allows solid wall structures to breathe and therefore to facilitate the evaporation of moisture. Modern paints seal these walls, giving rise to damp problems and damage to lime plasters and renders. The application of a limewash can also rejuvenate old lime renders.

Left: Internal limewash at Bellinter House, County Meath

Right: The basic ingredients for limewash: lime putty and water

MIXES

Q. What is a common mix for limewashing?

Common styles of brushes used for limewashing

A. The most efficient mixes are generally weak. Multiple thin coats are better than fewer coats of a thicker mix. In general, coats are about one part lime putty to three to six parts water. Experienced practitioners can tell by the consistency, appearance and how the limewash behaves during application whether the mix is right or not.

It may be tempting to apply a much thicker limewash mix in order to get the job done more quickly. This thicker coat dries too quickly before carbonating; it cracks and flakes off. A series of thinner limewash coats is best, each carbonating before the succeeding coat is applied.

PRE-WETTING

Q. Should I pre-wet surfaces with water before and during limewashing?

A. Pre-wetting of background surfaces is essential. It prevents newly applied limewash from drying out too quickly and failing. A small amount of lime putty can be added to the water in the pre-wetting process, particularly when the background surface is highly porous and friable and needs to be consolidated first.

After each of the coats (normally about five) of limewash has been applied, it should be prevented from drying out with an occasional fine mist spray of water. Damp hessian should be hung in front, a distance away from the wall, to reduce drying out. Corners of buildings and the reveals of open windows and doors may dry out more quickly and may therefore require more frequent wetting than other areas.

Just before applying the next coat of limewash, the preceding coat should be lightly sprayed with water.

PREPARATION

Q. How should I prepare a rubble stone wall before limewashing?

A. The wall may need to be pointed first. Holes, cracks, crevices, loose pinnings and missing stones should be filled or consolidated before limewashing. Otherwise, they will be very noticeable and throw dark shadows. Also, rough projecting areas of poor pointing carried out previously may need to be removed first or they will show through the limewash.

Limewash is not successful when applied over impermeable background materials such as sand and cement, concrete, oil-based paints or emulsions.

BUILDING ELEMENTS

Q. What building elements reduce the effectiveness and longevity of limewash?

A. Here are a number of possibilities.

- Faulty rainwater goods such as gutters and downpipes discharge water onto wall surfaces and remove limewash. They also cause dampness in the structure. Where large volumes of water are discharging onto a wall, this not only removes limewash but also bedding mortar between stones, causing dampness and plant growth. In severe cases, it can also undermine the base of walls.

- Window cills, capping, barge stones and spouts that do not project sufficiently, or that do not have throats or drips on their undersides, can cause problems with concentrations of water running down walls.

- Parapet walls can also present challenges. They generally have to cope with more severe exposure to wind and rain, and may require more frequent applications than less exposed walls.

- Retaining walls are continually damp and are unlikely to be limewashed successfully.

- The base of a wall is particularly vulnerable from splashing and rising damp with salts. Careful detailing is required here. A good principle is that buildings should have *a good hat and a good pair of boots*. This means a sound, overhanging roof for shelter and a plinth (if appropriate) with a well-drained edge strip at the base of the wall.

QUALITY LIMEWASH

Q. What are the key points in producing a quality limewash?

A. Consider these points in producing a quality limewash.

1. **Location**
    Walls facing prevailing wind and rain can result in a short life-span for limewash, while sheltered walls will provide better conditions for a longer life.

2. **Substrate and its Preparation**
    The nature of the substrate itself is important. A permeable substrate is best, allowing the limewash to soak in and carbonate. If impermeable, the limewash merely lies on the surface and is liable to be washed off easily by rain.
        The preparation of the substrate (pointing, daubing out, filling of holes etc.) needs to be done with an appropriate matching lime mortar. A brush may be useful for working and texturing mortar repairs flush to the face of the stone. Smooth, tight finishes from steel tools should be avoided.

3. **Materials**
    Lime putty is normally used for limewashing. Water is added to this, along with pigments, if required.

Existing wall being prepared for limewashing. Kilmallock, County Limerick

Small holes in existing wall filled with lime mortar prior to finishing limewash. Kilmallock, County Limerick

Secondary mixing of limewash by hand after earlier mechanical mixing on the same day. Yellow Ochre has been added to the tub on the left

In the past, the final coat may have had the addition of tallow (clarified animal fat), raw linseed oil or other such materials to improve the degree of water resistance and prevent uneven colouring when the wall is wet. They reduce the ability of the wall to breathe and may cause discolouration over time, preventing the addition of further limewashes for a number of years. Avoid their use, if possible.

4. Material Preparation

Lime putty and water are mixed together thoroughly. Preferably, this is done using a long electric drill mixing attachment, taking due care when working with electricity and water. Quite a large amount of limewash

Raw Sienna earth pigment

Application of limewash with a
traditional limewashing brush

can be prepared at one time in a large three-quarter full domestic waste disposal bin. Concentration on thorough mixing helps produce a quality finish.

Colour may be applied to all five coats, but is sometimes only added to the final coat or two. Earth pigments such as *Venetian Red, Raw Sienna, Burnt Sienna* and *Yellow Ochre* are traditionally used. These can first be added to warm water and mixed thoroughly before adding to the limewash and mixing again. Throughout the day, occasional further mixing will be required.

Keep a small quantity of coloured limewash from the first mix. Add this to the top of any new mix; check for colour match and adjust the new mix accordingly. It is difficult, if not impossible, to achieve a totally uniform colour over an entire wall but this subtle variation is one of the beauties of limewash. Another is that you can build up numerous layers over time until a thick protective coating is achieved. Many old buildings have multiple layers of limewash which add to their beauty and resistance to weather.

5. **Weather**

Rain, frost and very warm, sunny weather can affect the quality of the work in progress. A mild, overcast, damp day offers the best conditions for applying limewash.

6. **Application**

The application should be in multiple coats – five or more is normal on new work.

These coats should be thin, as thick coats will later flake or dust. Application by hand using a limewash brush is best. The wet bottom edge of each coat should be followed down the wall with the brush because on drying, these runs are sometimes visible under later coats.

On wet dash, particularly new wet dash, it may be necessary to throw or flick limewash from the brush into holes and crevices. As each coat dries, it should be burnished with the brush in all directions.

While it is tempting to add further coats on the same day, applying one coat only per day is recommended. The contents of the bucket should be stirred regularly.

When first applied to an external surface, limewash may suffer a little from weathering in its first year. For this reason, it is advisable to apply further coats after a year. The limewash should have a reasonable life after that, possibly up to five years, depending on its exposure to the elements.

7. Protection and After-Care

Keep newly applied limewash damp by spraying with a fine mist of water. The limewash should not be allowed to dry out during the course of the day of application. The following day, it should be sprayed lightly again and another coat of limewash applied. This coat is then kept damp throughout the day... and so on until the final coat of limewash is applied.

Newly applied limewash needs to be shaded from the sun and protected from drying winds and excessive rain. Damp hessian secured a short distance from the face of the wall works best.

Each coat of limewash is sprayed throughout the day with a fine mist of water to prevent rapid drying out. Damp hessian covering provides a moist environment and shade

Limewash on old masonry that has been pinned and flush pointed at the Church of St Carthage, Rahan, County Offaly

Detail of limewash on the Church of St Carthage

LIMEWASH – DETERIORATION

Q. The old limewash on my outbuildings is flaked, cracked and falling off. Is this normal?

A. Limewash applied in thick coats is liable to shrink, crack and flake. Over time, it will fail. A thick coat was commonly applied to achieve the required finish in one go rather than applying multiple thin coats.

The removal of existing limewashes should be done only as a last resort. They clearly record, layer-by-layer, every coat of limewash that may go back in time to when the first coat of limewash was applied to the building. Usually, old pigments are easily discernible and can be used as a resource when deciding what colours to use in future coats.

Before commencing, brush the wall down to remove loose, flaking material. Retain as much of the old limewash as possible. New coats of limewash on top of old can sometimes help consolidate earlier coats.

RENDER LIMEWASH

Q. Should a newly applied lime render be limewashed?

A. Lime renders are generally rejuvenated, preserved and have a longer life if a limewash is applied. It should be seen as an integral part of working with such renders. Limewash should also be renewed every so often.

FERROUS SULPHATE LIMEWASH

Q. I want to recreate the yellow/orange limewash on the external wall of a building. Apparently the pigment used was in green crystal form. What type of pigment was this?

A. It is likely that this colour was created by adding ferrous sulphate to lime putty and water. This is available in the form of a green crystal and is relatively cheap. In Ireland, it was used commonly in the nineteenth century. It is still used successfully in Scandinavia.

Ferrous sulphate crystals are mixed in warm water first, then with lime putty and water. When applied, it goes on green

A ferrous sulphate limewash applied to a lime wet dash showing some loss of colour over the years and now ready for another coat or two

Venetian Red earth pigment

but quickly turns to yellow/orange. A consistent and more attractive colour is dependent on using a more refined grade of ferrous sulphate rather than the cheaper farm-store variety.

VENETIAN RED LIMEWASH

Q. Traces of an old limewash are blood-red but faded to pink in places where it is exposed to the weather. How can this be replicated?

A. This was probably an earth pigment called Venetian Red, commonly used in limewashes in the eighteenth and nineteenth centuries. It was also used in colour washes applied to brick façades in Dublin to produce an even-coloured brick.

Venetian Red is first mixed in warm water and then with lime putty and water. The amount added to the mix determines the depth of the finished colour.

One of the most popular of the colours, Venetian Red is a strong colour requiring less pigment than some lighter ones. It is reminiscent of what was seen in Italy and elsewhere on the Grand Tour enjoyed by wealthy eighteenth and nineteenth-century travellers.

WHITE CEMENT

Q. For limewashing, it is common practice in my area to add white cement to hydrated calcium lime and water. This is done to ensure that the limewash stays on the wall. Is there anything wrong in doing this?

A. *Never add cement to limewash.* Limewash allows solid stone and mud walls to breathe. Adding cement negates this advantage. Instead of using bagged hydrated lime, use lime putty instead.

BACKPACK SPRAYER

Q. Should I use a backpack sprayer to apply limewash?

A. It is best to apply limewash with a brush. It is possible, however, to use a backpack or garden sprayer as it can be very useful in getting lime into holes, cracks and crevices that otherwise would be difficult to reach. Newly applied wet dashes are rather sharp at first, making it difficult to apply limewash with a brush. Over time, additional limewashes and the weather create an attractive, undulating finish that is much easier to limewash using a brush.

As the filters in the sprayers are likely to clog very easily, these should be removed to make the process a little easier. Clogging may still remain a problem, however.

Even with these problems, it is possible to use a sprayer and to unclog it as you go. As with brush application, limewash should be applied in multiple light coats. The temptation is to try and achieve the final finish in one coat – this should be avoided.

Finish all coats by burnishing with a brush after applying limewash with the sprayer.

Natural hydraulic lime render. Oriel House, Ballincollig, County Cork

PART

3

THE SPECIFIER

INTRODUCTION

This section of the book will be of assistance to the architect, engineer, building's surveyor, quantity surveyor, clerk of works and conservation officer, all of whom specify, design, quantify, control, advise or encourage people to use lime for the repair of old buildings or structures.

Writing the specification for a traditional heritage building needs to encompass more than just reference to EN standards. It should reflect a thorough understanding of the building, material and processes in order for a contractor to feel confident in pricing and in carrying out the work. An architect or engineer who specialises in architectural conservation is best employed for the task. Skilled craftspeople will be necessary to carry out the work, and this should be noted in the specification.

It is the specifier who sets the scene – what is to be done, what is to be left alone, what materials are to be used,

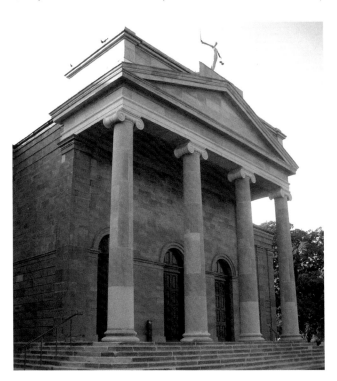

Delaminated sandstone columns repaired using a patented repair mortar and joints pointed using a lime mortar. Courthouse, Nenagh, County Tipperary

Above left: Mortar loss between stones is evident, but very careful consideration is needed before any intervention on a structure as precious as this. Temple na Hoe, Ardfert, County Kerry

Above right: Seventeenth-century brickwork, a rarity in Ireland, undergoing conservation and repair using lime. In this instance, a minimal intervention approach is called for; the building will still be a ruin but the decay process will be slowed down. Jigginstown House, Naas, County Kildare

what standards are to be established and maintained. In discussion with the relevant authorities, the building owner and the contractor, it is the specifier who establishes a philosophy of approach that will govern all decisions, actions and outcomes for the duration of the project.

The following are some of the underlying principles that should govern interventions to old buildings.

■ **Minimum intervention** means repairing only what needs to be repaired. For example, to point only what needs to be pointed – selective rather than global pointing. In this way, both the historic fabric and the embedded energy of the building are preserved.

■ **Repair like-with-like** means matching mortars or coatings to the original in order to maintain similar vapour permeability and flexural strength. In other words, any new intervention will have behavioural characteristics similar to the old in terms of moisture evaporation and movement. To achieve this, analysis of the existing mortar must be carried out. If the existing mortar has failed badly because it was inappropriate to begin with – for example, high calcium lime mortar without the addition of a pozzolan used at the base of

Minimum intervention and like-with-like. A minimum amount of deteriorated stone, c. 20 mm deep, was removed from column drums. This was replaced using a repair mortar with characteristics similar to the existing stone. The joints were then pointed with a matching lime mortar

piers and abutments in bridges – a performance-based mortar for that particular situation should be used.

■ **Reversibility**. One of the main advantages of most lime mortars and coatings is that they are relatively soft and capable of being reversed, without damage to the original building fabric. If necessary, the work done now can be undone at any time in the future. Reversibility also means that when a lime-built building is being demolished, materials like brick and stone can be recycled.

MORTAR TYPES AND MIXES

Q. Our architectural practice is about to concentrate on the repair of older buildings. What types of mortars and mixes might we have to contend with, both as-found in the building and as repair mortars?

A. A whole range of mortars may be found in an old building, varying from mud to lime to later repairs with modern cement mortars. The table (pages 162–63) gives a brief outline of past, present and future mortars likely to be encountered in Ireland. It does not include all the variances currently used in repair mortars, some of which are purposely designed for very specific repair situations.

Mixes vary a great deal, and the mixes shown here are only general. Again, advice should be sought before proceeding with any work. Mix ratios such as 1:3 always show the binder first and the aggregate last. Therefore 1:3 in a lime mortar mix indicates 1 part lime to 3 parts aggregate. On-site mixes are usually gauged by volume, although it is more accurate to gauge by weight when using hydrates such as natural hydraulic lime and sand mixes. Water used should be drinkable and conform to EN 12518.

The table on pages 162–63 is an outline only and should not be used without expert advice.

Although typical mixes are given here, it is important to analyse and match mortars in older buildings.

Left: The Browne Clayton Column,
Carrigbyrne, County Wexford, built
in 1839 and struck by lightning in
1994, prior to repairs

Right: The Browne Clayton Column
after repair

Below: The top third of this
monumental Corinthian column,
with its internal staircase, was
rebuilt using natural hydraulic lime
mortar in 2002-3. Some of the
individual cantilevered stones (the
volutes) weigh 1.5 tons

Mortar Type	Binder/s
Lime putty mortar EN 459: Part 1:2001	High calcium lime $Ca(OH)_2$ Variously called non-hydraulic lime, fat lime, CL90, air lime.
Hot lime mortar	High calcium lime as before but as a quicklime, calcium oxide, CaO (EN 459: Part 1:2001). Also called lump lime.
Pozzolanic lime mortar	Pozzolan added to a high calcium lime or more rarely to a natural hydraulic lime.
Hydraulic Limes **Natural hydraulic lime** EN 459: Part 1: 2001	NHL does not contain any performance-enhancing additives. Its natural mineralogy imparts a hydraulic set.
Formulated lime EN 459 revision due c. 2010)	NHL and/or calcium lime plus added hydraulic lime or pozzolanic materials including cement. Manufacturer must disclose contents.
Hydraulic lime EN 459 revision due c. 2010)	Lime and other materials such as cement, blast furnace slag, fly ash, limestone filler. Manufacturer not obliged to disclose contents.
Natural cement (Prompt by Vicat)	Natural cement obtained by burning an argillaceous limestone followed by grinding.
Mud mortar	Clay
Portland cement (EN 197-1) and lime (EN 459) mortar	Cement is the principal binder with the lime used to increase workability.

Where used	Typical mixes
For repair of old buildings, mainly internal use. External use limited, except generally with a pozzolan added.	Mixed with sand, 1:1.5, 1:2, 1:2.5 and 1:3 for backing coats and mortars (plus animal hair for plastering). 1:1 to 3:2 for finish coats. Mixes can vary widely.
Traditionally used in stonemasonry. But rarely if ever used today. Seen in historic renders, brick mortars, plaster backing coats. A pozzolan sometimes added. Danger of expansion and popping, particularly in plaster coats.	Mixed with sand approximately 1:3 plus water produces a boiling-hot mortar that was either used hot or soured out and later used cold. Quicklime was sometimes hydrated just before mixing with sand. Safety issues.
Building mortars, plasters and renders for the repair of old buildings.	Various pozzolans such as brick dust and calcined and finely ground clay available. Added as a % of lime in the lime and sand mortar mix. Check with the manufacturers re % quantities to use.
Mortars, renders, plasters, limecrete and hempcrete in repair of old buildings and in new-build. The most commonly used hydraulic lime.	In 3 grades: NHL2, NHL3.5 and NHL5. Mixed with sand at 1:1.5, 1:2, 1:2.5 and 1:3. Also mixed with bio aggregates.
Likely to be used mainly in new-build.	In 3 grades: FL2, FL3.5 and FL5. Mixed with sand, 1:1.5, 1:2, 1:2.5 and 1:3.
Likely to be used mainly in new build.	In 3 grades: FL2, FL3.5 and FL5. Mixed with sand, 1:1.5, 1:2, 1:2.5 and 1:3.
Engineering works, below ground or water, decorative castings, external run *in situ* mouldings etc.	Mixed with sand, sometimes with a retarder (citric acid) to slow setting speed. Mixes 1:2, 1:3 and 1:4.
Traditional mortars, renders, plasters, walls, floors, fire breasts, firebacks. Puddling over bridge arch barrels, linings to ponds and canals. Repair of old buildings; some new-build.	Mud (subsoil with clay, silt, sand, sometimes gravel). Used as-found or with addition of sand; also animal manure, straw and lime.
New-build mortars, plasters, renders, concretes. Not for repair of old buildings.	Various mixes such as 1:1:6 and 1:2:9 (cement, lime and sand).

The compressive strengths in N/mm² at 28 days for NHL2, NHL3.5 and NHL5 are determined in accordance with EN 459-2-2001. NHL1 is a new addition (2007) to the range that is not covered in this standard.

NHL1 (new) — check with the manufacturer
NHL2 ≥2 to ≤7
NHL3.5 ≥3.5 to ≤10
NHL5 ≥5 to ≤15

It can be seen from the above that compression strengths can vary widely from initial to final testing stages.

Unlike Portland cement, the testing of natural hydraulic lime at 28 days does not reflect closely its final compressive strength; this is more accurately done at 90 days.

Individual manufacturer's product data sheets and current European standards should be consulted for further information.

BRIDGE REPAIR

Q. Our engineering practice has been given the brief to produce a specification for the repair of a series of nineteenth-century stone bridges. They are all built in hard blue/grey carboniferous limestone. Cut stone occurs only in the arches, copings (partly missing) and in the quoins on the face of piers and abutments. The remaining stone is rubble limestone with varying joint sizes.

From examination of these bridges, they appear to have developed similar deterioration characteristics. Partial loss of original lime mortar and stone from the underside of the bridges is widespread. Severe loss of mortar is observed at the bases of some piers and abutments, as well as on bridge faces under parapet drainage holes. Occasionally, there are cracks in wing walls, and cutwaters are sometimes separated from piers.

Rather than using cement-based mortars and concretes for repair, we are considering the feasibility of using lime mortar but are concerned that it may not be suitable for the job.

A. Most of our old stone bridges, and there are over 20,000 of them in Ireland, were built using lime mortar.

Mud as a mortar and as a *puddle* (tempered mud) over *arch barrels* is also found. There are even a few dry stone bridges.

These bridges have served us well, not only because masonry is one of the most successful means of creating a robust structural span but also because of the mortar that was used.

As in all masonry structures, lime mortar has two great advantages. The first is that it allows movement to take place within the joints without cracking the stones. The movement in a bridge is caused by both thermal expansion and contraction, and any live-loading created by traffic. The second advantage is that lime mortar creates a permeable structure that allows the water, which inevitably gets in, to get back out again.

A typical bridge can be divided into the following elements:

- foundations
- abutments
- piers and cutwaters
- wing walls
- arch barrels
- spandrel walls
- spandrel fill (fill over arch barrels)
- parapets
- surfacing

Each bridge and all of these elements must be examined to determine what repairs, if any, are required. A lime mortar suitable for the repair of one element may not be appropriate for the repair of another. The choice of which lime mortar to use and where is critical.

Water is the prime deterioration agent of all stone structures. A bridge not only has water running at its base, but also water from above, and splashing from passing traffic. It is likely that water has washed out the original lime mortar, washed fines from the spandrel fill, undermined piers and abutments, and removed mortar from between the stones forming the arch barrel.

The lime mortars that should have been used are either hydraulic or pozzolanic mix, but this was not always the case. Unfortunately, high calcium/non-hydraulic lime seems to have been used even below the water line. To prevent the wash-out of lime mortars in arch barrels, a

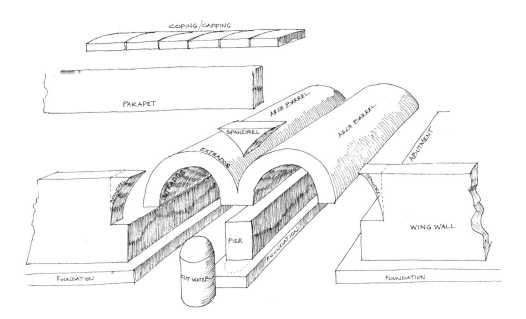

Opposite, top: Typical bridge elements

Opposite, bottom: Crack in the wing wall of bridge. County Limerick

Right: A replacement voussoir is being checked for fit before insertion using lime mortar. The voussoir will be finally fixed in position using pinning stones. The inverted 'V' shape of the voussoir reflects the shape of the hole, commonly found on the intrados of bridge arch barrels

cover of well-tempered puddle (mud) was sometimes used over the top of the barrel.

The original mortars used are not responsible for all of the faults. Not properly bonding the arches on the face of the bridge to the arch barrel or cutwaters to piers are faults sometimes found in the original construction. Cracks in wing walls may have occurred from settlement due to building on unconsolidated fill or too-shallow foundations.

Voussoirs (arch stones) missing from the underside of arch barrels are sometimes due to a loss of mortar, but also because of the way in which some of these stones were laid in the original construction.

Repairs below the water line will require a lime mortar with a greater hydraulicity than above the water line. Similarly, the inside of parapets may suffer from the water splash of passing traffic and the tops of parapets (without copings) from falling rain.

A lack of positive drainage (kerbs, gullies, longitudinal and transverse falls) can lead not only to vegetation growth in verges and wash-out of fines from the spandrel fill, but also to the loss of lime mortar from the arch barrel. Positive drainage on bridges will help to reduce these effects.

Pointing by itself is not always adequate. Grouting of piers, abutments and other elements may be necessary due

Repair work to bridge abutment
using stone and natural hydraulic
lime mortar. County Donegal

to mortar loss after construction. Deformation of the road
surface and potholes may indicate settlement of spandrel
fill or punching of voussoirs at the crown of arch barrels
from heavy traffic. It is often difficult to assess what voids
exist in spandrel fill and whether grouting, when used, is
successful in reaching these. Seek the expert advice of an
engineer experienced in traditional bridge repair.

Partial rebuild of the elements listed earlier may be
necessary if there is serious loss of original mortars and
stones in these elements. If plant growth, especially ivy, gets
established, it may result in the loosening, spalling and
jacking of stones; a partial dismantling and rebuild of wall
faces may be the only solution.

For isolated holes below the water line that need to be
repaired quickly, a fast-setting natural cement may be
considered. However, it is generally best to restrict the use
of natural cement to the minimum.

Bridges provide a habitat for many protected species,
particularly bats, birds, fish, otters and insects. A favoured
habitat for bats is under bridges. As they are protected
species, professional advice and permission must be sought
before commencing any work on a bridge. If bats are found,
the work can take place during unrestricted periods,
avoiding disturbance during breeding or hibernation.

Lime will also kill fish if it gets into water in any quantity. Mortar droppings from hand pointing and from rebound which may result from mechanical pointing must be prevented from entering the watercourse. Boarded scaffolding with plastic sheeting is usually a sufficient precaution in most cases.

The National Roads Authority Bridge Management Group (Republic of Ireland) have experimented, tested and run trials on using lime mortars to repair bridges. They have produced a specification for undertaking repointing using lime mortar.

If contractors do not have direct experience of using lime mortar, they should receive training in this type of work before being considered.

TOWER HOUSE

Q. Our client owns and lives in a sixteenth-century tower house, with a fifteenth-century great hall attached. As architects, we have been commissioned to draw up a specification for its repair. What concerns us most is that the amount of water seeping through the walls makes the building habitable only in the summer months. There are open joints between the stones on all external façades. Here and there are traces of what looks like an old wet dash rendering, but very little of this remains. With walls of considerable thickness, it is difficult to understand why so much water is getting in. What should we do?

A. Even though the walls are of considerable thickness, damp patches after rain will most likely occur first at the following openings through external walls:

- *Intrados* (underside or soffit) of arches
- The *soffit* (underside) of lintels
- *Jambs* (vertical and visible cross-section) of doors and windows

Prolonged and wind-blown rain will penetrate most walls some distance in from the face, with ingress usually seen first at the points mentioned.

Fire-damaged limestone displaying cracks and spalled faces on a sixteenth-century tower house in Dublin. Some of the joints were repointed with a hard lime mortar at some time in the past

Damp occurring on the inside faces of thick walls is another matter. Some of the reasons for this may be attributed to:

- Loss of mortar from joints in the outer face of the wall.
- Loss of external rendering.
- Wind-blown rain against the outer face of walls in very exposed locations.
- Cracked or fissured stones, either natural or as a result of a fire in the past.
- Exposed wall tops to parapets with missing crenellations and copings.
- External ground level higher than internal floor level.
- The overflow from water spouts and holes left through parapets being blown back against the building. The result is ingress of water, wash-out of mortar and plant growth.
- Water splash from spouts off pavement and passing traffic.
- *Allure stones* (the wall walk behind the parapet) are missing, cracked, off level or moved out of position, allowing ingress of water down through the top of the wall. A run-off of rain from a roof directly onto the allure stones can also result in a considerable concentration of water.

Allure stones (also known as water and saddle stones) behind a parapet wall on a roofless tower house

Hand-operated grout pump used to fill voids within the wall core

- Poorly constructed hearting in the centre of the wall.
- Loss of mortar from the centre of the wall.
- Water reservoirs within walls that leak water over a prolonged period.
- Previous attempt at external pointing using sand and cement which is now trapping water and preventing evaporation.
- Condensation, confused as ingress of rain, caused by temperature change; infrequent heating combined with poor ventilation.

Not all of these problems can be rectified, but those that can are likely to involve the use of lime, such as:

- Grouting
- Pointing
- Wet dashing

Grouting

Seek expert advice to ascertain whether a wall requires grouting or not – there may be structural implications involved in adding weight to a structure. Permeability and flexibility of the structure will be negatively affected if the lime grouts are too strong. Limes used in grouting need to be hydraulic in order to set where there is limited access to air. A 'free lime' content is generally considered beneficial in grouts used in structures above ground. Free lime can move with the aid of moisture to small voids, cracks and cavities where it can then harden. The addition of Bentonite (a type of clay) will increase the flow characteristics of the mortar.

Mortar is either injected under pressure or gravity-fed through a tube into the wall. Advice should be sought regarding the appropriateness of either method.

If an entire wall is to be grouted, it is normal to start about a metre from the base of the wall. The wall should be flushed with water first in order to clear out loose debris and dust. When dried to a damp state, the wall is ready to commence grouting.

One of the following methods is used:

- Plug the open joints between stones with puddle clay or hessian before grouting. After grouting,

Drill holes plugged with hessian after lime mortar grouting of wall core at Ardfert Cathedral, County Kerry

remove the puddle clay or hessian to check visually whether the grouting was successful. When the grout has set, point the wall

or

■ Point the wall and then grout.

Puddle clay is useful for plugging any joints that leak, as well as for creating cups on the wall face into which grout is poured when gravity grouting. It is worth noting that the weight of uncured liquid grout within a wall can blow it apart if too high a lift is carried out at once.

When the grout has set, the next one-metre lift follows, and so on until the wall is completed.

Pointing

Pointing is described under 'Pointing Rubble Stone Buildings', page 66.

Pointing may be all that is required, in addition to dealing with exposed wall tops etc. If not, the option of using a wet dash has to be considered.

Wet Dashing

Wet dashing is also described under 'Wet Dashing', page 136.

The application of a lime wet dash to medieval structures is sometimes controversial. The controversy is centred mostly on aesthetics and the shock of seeing a much-loved stone building disappear behind a lime render. The reality is that almost all rubble stone buildings were originally rendered but have lost these lime coatings over time. Close examination of stone buildings will often reveal traces of these original coatings where they have survived in sheltered areas. A common misunderstanding is that the renders were applied at a much later date, perhaps during the eighteenth and nineteenth centuries as part of a popular style. Visual evidence and scientific analysis have proved that many of these coatings are original features of the ancient buildings on which they are found.

If it is decided to proceed, samples of the existing wet dash should be analysed to determine the colour, shape, size, grading and geology of the sand/aggregate, and the lime type and mix, so that an appropriate mortar mix can be designed as a replacement. The sands and aggregates used in the past were usually sourced from the local area, as they should be if possible with any proposed repair work.

Preliminary preparation of wall surfaces, treatment and removal of vegetation, pointing etc. must be done, with the method and style of application determined and carefully controlled.

Of course expert advice must be sought and permission obtained from relevant authorities before any intervention to the tower house commences.

Condensation

Finally, extreme condensation in these structures is sometimes evident on the inside faces of external walls. This is often wrongly diagnosed as moisture penetration due to driving rain or rising damp. Thick, solid stone walls adjust slowly to changes in temperature. Hard, smooth, dense stones like some Irish limestones are particularly vulnerable to showing condensation on their faces. Condensation can be reduced by ventilation and heat. A constant low level of heat throughout the building works best. Internal lime plasters and/or limewashes can also help to alleviate the problem.

Wet dash using a high calcium lime and a pozzolan at Rathfarnham Castle, County Dublin

Above left: Projecting eavestone with spike and gutter bracket fixed underneath. The bottom slates are often fixed to timbers set in lime mortar on the wall top. The undersides of slates are parged/torched with lime mortar

Above right: Eavestones , normally laid as a single course, here laid in multiple courses. County Donegal

EAVESTONES AND ROOFS

Q. My client owns an old roofless stone barn built in lime mortar. It has two gables and two long side walls. The side walls carried the roof and have the partial remains of a projecting stone, with a loose lime-mortared sloped finish on top. The wall tops are to be repaired in lime mortar in preparation for the construction of a new roof. This type of detail appears common on similar buildings in the surrounding area, but as an architect practising in new-build, it is new to me.

A. Today, many old buildings are being re-roofed and traditional details are being lost. The prefabricated roof trusses, fibre cement artificial slates, plastic gutters and downpipes all replace what was there before. The projecting stone mentioned is an eavestone or eaves course which is often wrongly replaced with a plastic soffit. The old details are neither understood nor are they being replicated. Here is a great opportunity to reinstate them.

Consider the following procedure.
- Remove loose stones from the sloped wall top down to the projecting eavestone.
- The eavestone is usually about 75 mm thick and projects about 75 mm beyond the external wall face. The first slates of the roof usually rest on the top projecting arris (edge) of this stone. Eavestones are common on older buildings in most parts of the country. Eavestones rarely

Slates being laid in lime mortar on a wall top. Buncrana, County Donegal

covered the complete wall top. Any replacement stones should be of the same type and thickness. Lay the stones in lime mortar with the same projection beyond the external wall face.

- The inside face of the wall should now be built up from the level of the eavestone to receive the wall plate for the rafters. The rafters often stop at the wall plate and do not cover the wall top.
- The roof is now constructed.
- The wall tops should be sloped at the same pitch as the roof, in line with the top of the rafters and finishing just short of the outer arris of the eavestone.
- Battens are usually bedded in lime mortar in the sloped wall top for fixing the slating battens. At other times, the slates are simply bedded in lime mortar without fixing to battens. Any wood used for battens should be dovetailed in section and treated to resist rot. These battens can be wider and deeper than normal slating battens. Above

Lime torching or parging to the underside of the slates at Russborough, County Wicklow. These 250-year-old slates are supported on timber pegs

this, a breathable felt/membrane is fixed to the top of the rafters and taken over the wall top.

- The roof is battened and slated.
- Gutters were not always used, and many roofs were gutterless. If gutters are to be fixed, they should ideally be of cast iron. Gutters sit in brackets that are fixed to a spike driven under the eavestone. Downpipes of cast iron, usually of quite small diameter, were traditionally used to carry the water from the gutters to the ground. A gulley should collect this water at ground level and take it away in a drain below ground level.
- The undersides of roof slates were traditionally *torched* (plastered) with lime, sand and animal hair. If you have used a modern, breathable felt/membrane over the rafters and under the slates, it will not be possible to do this.

REPLACEMENT ASHLAR

Q. Near the base or plinth of a granite ashlar building, there are signs of severe decay, with the stone de-fragmenting and falling to the pavement—like coarse granite sand. There is also localised cracking and spalling. It is intended to remove the worst of the damaged stones and replace these with new stone. A matching lime mortar is to be used throughout for joints, which are about 6 mm wide. The original stones are quite large and we have some concern about removing them.

A. A specialist contractor will be required to carry out this work.

Although the ashlar stones look rather large, they are rarely that thick. They were sometimes used as a thin façade or veneer against a thicker walled rubble stone or brick building.

In the Dublin area, older granite buildings were sometimes built with granite quarried from the surface of the land. This granite had been there since the last ice age, 10,000 years ago. Through weathering over such a long period, its natural iron content oxidised, expanded and made the stone relatively soft and easier to work.

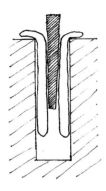

Above left: Plug and feathers used to split stone

Above right: A masonry handsaw being used to remove lime mortar bed joints in granite ashlar stone. St George's Church, Temple Street, Dublin

Right: Replacement of granite ashlar stones using a lime mortar. St George's Church, Temple Street, Dublin

Unfortunately, it also resulted in the early decay of the stone. New granite available today is quarried deeper and is of a higher quality.

Wrought iron cramps or 'dogs' were used to connect one stone to the other. These were usually set in lead. Again it is oxidation, in this case of wrought iron, that is causing problems; this manifests itself as spalling and cracking near where these cramps are fixed.

If a course of stones is to be removed, the first stone is usually the most difficult. If the original lime mortars are relatively soft, a masonry handsaw can be used to remove the joints. *Plugs and feathers* (a steel wedge set between two steel shims) inserted in drill holes may be used to split the stone for easy removal. The stones overhead are propped, as necessary, with heavy timbers to prevent collapse.

The next stone is removed similarly. After the removal of two stones, a replacement stone is inserted on a bed of lime mortar. The top bed is pointed and timber wedges inserted in case of overhead movement before the mortar has set.

To prevent damage to the existing stones that are to remain in place, avoid disc-cutting machinery as far as possible,

GEORGIAN BRICK

Q. The Georgian house for which I am writing a specification has damaged brick on the wall face. At the window reveals, it is also noticeable that the bricks on the external wall face have moved outwards, away from the inner wall face. The modern sand and cement pointing has failed, and is cracking and falling out of the joints, but the original lime mortar bedding behind this is in good condition. My concern is that the cement mortar is causing damage and that part of the external brick façade may collapse.

A. Damage to the faces of clay brick may be the result of any of the following:

- Sand and cement pointing has prevented drying out and is causing subsequent damage from frost.
- Soluble salts in the cement re-crystallising in the pores of the bricks.
- Some of the original bricks may be under-fired and susceptible to general weathering, frost and salts. These are usually the lighter-coloured bricks.
- Leaking downpipes and gutters may be washing out mortar joints, causing plant growth and brick decay from wet/dry cycles and frost.
- Damage to bricks on parapet walls is common because of their exposure to the elements from both sides. Inadequate cappings to wall tops and poor rainwater disposal and run-off details, or a build-up of debris between the parapet and the roof behind, also lead to the problems you describe.

The outward movement of the wall at the window reveals may be the result of *snapped headers*. Flemish bond, the most

Spalled eighteenth-century brickwork in Dublin

A brick wall under repair. The external face bricks have been removed to show how it was inadequately bonded to the stock bricks behind. St Stephen's Green, Dublin

common bond found in these Georgian houses, consists of alternate stretchers and headers showing in every course. Sometimes the header brick, instead of being bonded with its length into the thickness of the wall, was snapped in half to save on facing bricks (a snapped brick would produce two headers instead of one). This practice resulted in a weakened, less effective bond.

A cheaper stock brick was usually used behind the facing brick, thus forming the bulk of the wall. The facing and stock bricks were sometimes different sizes and bedded to different heights. This made the bonding of the face brick to the stock brick difficult, but not impossible. Sometimes the bonding was simply ignored.

The result can be an outer leaf of brick about 100-112 mm thick rising a considerable height, with inadequate bonding to the wall behind. It is likely that each wall was built separately rather than in tandem, as good practice dictates.

A different mortar may have been used in each wall, high calcium lime mortar internally and a pozzolanic or hydraulic lime mortar externally.

A structural engineer specialising in conservation should be consulted. There are patented means of drilling and inserting rods to tie both walls together at intervals that could be considered. However, epoxy resin rods change the permeable characteristic of solid brick walls, showing as unsightly damp spots when the wall has dried out after rain.

If a partial rebuilding is necessary, a lime mortar should be used to match the existing mortar in the wall. If the existing sand and cement mortar has to be removed, it must be done so as to avoid damage to the bricks. This is often a difficult activity and may require the input of a specialist contractor. The joint finish should replicate an original finish if one can be found. If not, neighbouring houses of similar age and detail, with original joint finishes surviving, should be examined and copied. Tuck pointing (see page 98) should not be applied to brick buildings that show no evidence that it was used previously. All mortars should be weaker and more permeable than the brick.

Bricks showing evidence of damage to their faces can sometimes be removed, reversed and re-inserted into the same position in the wall.

It is important that replacement bricks not only match the original bricks in colour but also permeability, size and texture. Generally, local salvaged bricks will be a better match than a newly manufactured copy.

IVY

Q. We want to repair an old stone building by replicating the existing lime mortars for pointing, rendering and plastering. Before the work can commence, extensive ivy growth must be removed. The ivy has dislodged stones from the walls. There is extensive cracking, and barge stones from the gable have been lost. We are looking for information on how best to eradicate the ivy.

A. An informative booklet on the subject is 'Dealing with Vegetation on Historic Masonry Monuments' from the Environment and Heritage Service, Northern Ireland (now the Northern Ireland Environment Agency, NIEA).

In a very short number of years, and once ivy has taken hold, it can be very destructive to masonry structures, particularly those with open joints, loose stones and exposed wall tops. If ivy is left unchecked and allowed to

The damaging effect of ivy on old masonry structures at the north-west tower, Athenry

establish aerial roots in mortar joints, whole sections of wall faces may have to be taken down and rebuilt. Ashlar stonework, even with 3 mm wide joints, is vulnerable, as the ivy gets into the joints and behind the ashlar stone face.

Ivy growth can cause the following problems:

- Dislodgement of stones off their beds from expansion of ivy roots in the joints and behind the stones.
- Spalling of stone arrises at mortar joints from expansion of ivy roots and stems.
- Collapse of walls and wall faces from expansion of ivy and increased wind loading.
- Cracking in stones and walls from expansion of ivy roots and stems.
- Danger to the owners, public and workers involved in repair.

Don'ts

- Don't pull ivy off a building using ropes or other similar means.
- Don't remove ivy before treating it.
- Don't cut ivy flush to wall face and then point over.
- Don't rebuild walls leaving roots systems inside.
- Don't remove ivy if it is a habitat for protected flora and fauna.
- Don't remove ivy that is holding historic wall fabric in position without recording what exists first and being ready to commence conservation of the structure immediately.
- Don't leave ivy in position after killing because shrinkage of roots and stems can cause a wall to collapse.
- Don't remove ivy without a follow-up work programme of repair – otherwise, important details on the building may be lost. Cutting back the ivy may be necessary to reduce wind loading on the structure and allow the necessary work to be seen.

Ivy can be treated well in advance of commencing work.

- Treat the green leaves with an appropriate biocide.
- If the ivy is cut first, it must be followed by immediate (within 15 minutes) and appropriate chemical treatment with pastes and/or sprays. This is a specialist area because of health and safety issues, and the possible impact on surrounding flora and fauna, watercourses etc. The most up-to-date information on appropriate chemical use and safe application should be used.

HOT LIME MORTAR

Q. The original nineteenth-century specification for a train station specifies a mix of one part lime to three parts sand for the building of the brick and the stone. However, when we mix one part lime putty to three parts sand, we find that the result has far less lime than the original specification. By analysis, we have found that the original mix more closely approximates a 1:1.5 mix. Why would they have used twice the amount of lime to sand as was specified?

A. Many old specifications mention such mixes. Even simple observation shows that the lime content is higher than one part of lime putty to three parts of sand. One reason for this may be that they were mixing quicklime (calcium oxide) with sand in the proportion of 1:3. On slaking with water, quicklime produces approximately twice as much lime putty by volume as the original quicklime. Sometimes, old specifications and writings took it for granted that this was understood by everyone involved in building.

It is unlikely that the lime they used was a lime putty, and it is probable that they used quicklime. The quicklime would probably have been high calcium or fat lime, possibly used with a pozzolan. Or the quicklime used could have been feebly hydraulic. Quicklime was sometimes converted to a hydrate before being mixed immediately with the sand. This could have been used right away or allowed to sour out for an extended period before using.

Today, the use of hot lime mortar for the repair of buildings in Ireland is virtually unknown. There are also serious health and safety implications in preparing such mixes. However, it has been used elsewhere with success, and specialists may find that it is worth reviving for certain projects. If not mixed thoroughly, hot lime mortars may create later problems from expansion and jacking, particularly on thinner wall construction and near wall tops etc.

All workers involved in the use of hot lime mortar must be trained first, with proper safety procedures and personal protective equipment put in place. It is advisable that you have the existing lime mortar analysed in a laboratory specialising in such work before proceeding further.

FAMINE WALL

Q. As an architect representing a local community, I am responsible for the repair of an old Famine boundary wall that is a listed structure. It was built as part of a Famine relief programme and provided local employment in the mid nineteenth century. The wall is 600 mm thick, just over 2 m high, and about 1 km long. It is in reasonable condition, except for the top 300 mm or so that is loose, with some missing stones. The mortar between the stones is virtually gone and has been replaced by a dark, rich, loamy soil. No coping or capping exists on top of the wall, and there is no evidence that any ever existed. There is some pointing to be done here and there on both wall faces but otherwise the wall is sound. The stone is rounded glacial till, mostly hard limestone. What should I do?

A. The wall top probably requires rebuilding. This will involve the careful removal of the top 300 mm, which is best done by removing the stone in courses on each side of the wall. Place the top course furthest away from the wall, and as you work down the wall, place subsequent courses closer to the wall. The face of each stone should face away from the wall so that they can be placed directly back in position, thereby avoiding confusion about which way they should be turned. An even more accurate way of doing this

A plastic-covered frame is aligned to a grid marked on the wall. Each stone is then marked and numbered for dismantling before being rebuilt

is by drawing a grid on the wall and numbering and recording each stone before removal. Being a listed structure, you need to check what criteria you should follow.

Apart from the top 300 mm, it appears that the original mortar is still doing its job. It may be helpful to have the original mortar (near the base of the wall) analysed. It may be high calcium or feebly hydraulic. The replacement mortar for the top course and horizontal top surface/bed of the wall should be sufficiently hydraulic to reduce ingress of water down through the wall. This is especially so because there will be no coping or capping. At the same time, it should not be so radically different as to cause cracking and failure between the wall top and the wall underneath.

Match the sand in the repair mortar to the original.

With no coping on this wall, it is important to consolidate and point the top of the wall carefully. Hearting and pinnings should be packed with mortar into all open spaces and joints. Mortar joints should be compressed with flat jointing bars and all joints contoured to throw water off the wall top.

Pointing of the wall faces should be done selectively, working only where strictly necessary. The mortar for the vertical wall faces below the top 300 mm should match the existing mortar as closely as possible.

Let's assume that elsewhere on the wall, the original mortar joint finish is reasonably flush with the face of the

Rubble stone wall. The stones are irregular shapes, including round. What looks like a semi-rendered appearance is the result of flush joints and the expansion of hot lime mortar. This wall does not require repair

wall, showing exposed aggregate. If so, it is desirable to achieve the same on the new work. Beating with a stiff brush when the mortar is nearly hard ('Pointing Rubble Stone Buildings', page 66) is best. During the course of the work and for a period of time after the work is finished, the wall top should be protected from frost, wind, rain and sun with covers. To prevent the mortar from drying out, a damp hessian cover is advisable.

The new lime mortars will stand out for a number of years until weathering tones them down. No attempt should be made to colour the new mortar artificially in an attempt to match the existing mortar.

Finally, check the wall for any evidence of a lime wet dash. Many walls built of rubble stone were rendered with a wet dash both for aesthetic reasons and to prevent the loss of bedding mortars. If this is so, consideration should be given to its replacement.

POINT, REBUILD OR LEAVE ALONE?

Q. Our firm of architects is responsible for the repair of an old rubble stone warehouse which was built using lime mortar. The structure was partly pointed with sand and cement in the past. We must make decisions about where to point, rebuild and leave alone. Can you offer some advice?

The lime mortar is reasonably flush to the face of the wall and, although soft, has survived successfully, requiring no repair for two hundred years

A. The following information will help you make your decision.

When to point

Selective areas of wall showing:

- Excess mortar loss from joints.
- Loss of stone pinnings from larger mortar joints.
- Existing sand and cement pointing is causing problems, such as retention of water in the wall.
- Minor plant growth with roots tightly embedded in mortar joints.

When to rebuild

Selective areas of wall showing:

- Larger stones are loose and in danger of collapse.
- Isolated missing stones.
- Unstable sections of wall.
- Partial collapse of wall face.
- Destructive ivy and other plant growth has become deeply embedded in the mortar joints and behind the face of the wall, loosening and spalling stones with its powerful root system.

When to leave alone

- When there is an absence of any of the above, and when the mortar (although possibly soft) is still reasonably flush to the face of the wall.
- Remember to repair only what needs to be repaired and leave the rest alone.

When the scaffolding is erected, the walls of the warehouse should be inspected carefully from top to bottom. At this stage, decisions can be made about which areas to point, leave alone or rebuild.

Areas for pointing should be marked so that everyone involved is clear about what is to be done. A sample section should be pointed and finished for inspection before the work proceeds. Areas for pointing should be raked out first throughout the structure before any pointing commences.

Plant growth (such as Valerian, as shown here) on wall tops and wall faces is destructive, causing degradation of mortar joints and loosening of stones. Advice should be sought before removal of growth, as some plants may be protected species and/or may not be destructive to the wall

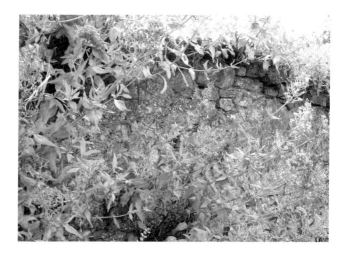

This is so the depth of the raking out can be inspected by the architect.

Areas of work requiring rebuilding should be decided upon and marked clearly. The same stone type should be laid in the same pattern, with similar joint sizes and placement of pinning stones. For stability, it is best if stones are always laid with their longest dimension running into the wall thickness.

Hearting stones should be laid without excess mortar, at the same time ensuring that they are not laid dry. (See 'Hearting' in Stonemasonry section of Practitioners, page 84.)

POINTING – QUALITY CONTROL ON SITE

Q. As a clerk of works, I have been appointed by the architect to oversee the quality of work on a conservation project that involves pointing a stone structure. What elements of the work that relate to lime should I be aware of?

A. Watch for a number of key points.

- Mix ratios are notorious for not being followed – even more so if there are several different mix ratios on the one site. It is one of the most important jobs to be carried out on-site but is often left to whomever happens to be available at the time. The verbal transfer of mix ratios from one operative to another is bound to come unstuck if not monitored continually. Post the mix ratios on a board in large letters so everyone can refer to it.

- Sand should be checked for size, grading, cleanliness, void space, colour, shape etc. If the void space is greater than allowed for in the mix design, and if there is no alternative but to use the sand, adjustments may have to be made to the mix with the agreement of the architect.
- Samples of materials such as sand should be kept from each delivery, as well as samples from pre-mixed mortars (if they are being used), pozzolans, hair, pigments etc.
- Water content in lime mixes requires careful control. Keeping the sand dry while in storage will alleviate radical change to water content.
- The use of a mortar mill, rather than the common cement mixer, is preferable for lime putty mixes. The mill physically works the lime putty and sand together rather than just turning it. Because of this beating action, it reduces the amount of water required to produce workable mortars, both for lime putty and hydrate mixes. It is important to set the roller at a sufficient height so as not to crush the sand. Paddle mixers are excellent too, and preferable to using the rotary drum cement mixer. The rotary drum cement mixer will mix lime hydrate, sand and water to an acceptable level for most types of work, but not when stiff, workable mixes are required or for lime putty mixes.
- Separate sample panels of all pointing, plastering, rendering or limewashing should always be provided and retained as a reference until the job is finished.
- A high calcium lime and sand mix can be mixed well in advance of using. It can be stored in a damp state under cover to keep it out of the air.

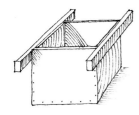

A traditional batching box that requires two people to lift

A plastic bucket used for batching, with the mortar mix clearly marked

This process should achieve a consistent lime:sand mix ratio, compared to mixing each mix as required. If a pozzolan is to be added to the mortar, it should be done just before use. Measurement by weight is preferred for natural hydraulic lime and sand mixes. This ensures accuracy and compliance with standards, although the less accurate method of measurement by volume is the most common. When measuring by volume, all materials should be measured in containers which have been marked and set aside for that purpose. These are called *batching boxes*. Small plastic buckets can also be used, but not shovels as they are too inaccurate.

■ Ensure that Portland cement is not being added to lime mixes. If used on-site for other purposes such as concrete floor slab or foundation, cement should be stored well away from where the lime is stored or mixed. Separate mixing machinery should also be used.

■ Pointing is carried out in stages, as described in 'Pointing Rubble Stone Buildings', page 66. To ensure quality of work, raking out should be inspected in stages before the pointing commences. This ensures that the raking is sufficiently deep and clean cut. Otherwise, it is difficult to ensure that new pointing is not being placed over existing joints that were not raked out sufficiently.

■ Pre-wet all the joints within a set area of work before pointing commences. Pre-wetting prevents the rapid drying out and consequent failure of new lime mortar. The amount of water used in pre-wetting is very much dependent on the weather and the porosity of the existing mortar and the stone. Just prior to applying the pointing mortar, the existing joints should be damp rather than wet. Pinnings should be damp before their insertion into larger mortar joints.

Joints raked out and pre-wetted before pointing

- It is essential to protect all work from frost, wind, rain and sun, and various types of temporary sheeting are used for this. Protection is essential during the course of the work and for a period of time after finishing. Allow for a couple of weeks. Cold weather will extend this period. (See 'Pointing Rubble Stone Buildings' in Practitioners – Stonemasonry for protection and curing.)
- To ensure proper curing of the mortar, wet hessian should be hung about 100 mm from the face of the work and kept damp throughout the curing period.
- Good organisation of the work activity is critical. The establishment of a good system of work is essential for safety, quality and productivity.

MORTAR QUANTITY

Q. We are building a wall using rounded field stones averaging 150 mm in size. In terms of quantity, how much lime mortar will it take, and is there any way to reduce this?

A. A wall composed of rounded field stones or glacial till, on both sides of the wall and as hearting in the centre, will require more mortar than a similar wall built with flat, square-shaped stones. Take care to fill the centre of the wall with as much stone as possible and as little mortar as possible. Ensuring that all stones are laid and jointed in mortar and not touching each other will require up to 50% mortar with rounded stones. In other words, the volume of such a wall may be as much as 50% stone and 50% mortar. The mortar volume in rubble stone walls generally varies from about 20% upwards.

Reducing mortar content in stone walls

- Carefully hearting the centre of the wall with stones graded from large to small will reduce the volume of mortar required and produce a better wall. Too often, mortar only is used for filling wall centres.
- Flat bedded stones require less mortar than rounded stones.
- An increase in stone size will reduce mortar volume, but if too large will also reduce production.
- Large cut stone, such as ashlar with 3 mm joint sizes, requires the least amount of mortar, although some traditional ashlar had only tight joints at the face of the wall with quite large joints behind.
- The rough shaping and dressing of stones with a hammer will result in smaller and more controlled joint sizes, therefore requiring less mortar. Cutting, particularly excess cutting, is not always practical, economical or even desirable.
- Pinning of larger mortar joints on beds and faces with small stones will reduce mortar volume.
- Management of materials at the purchase, storage, mixing, laying and protection stages avoids waste.
- Sand that is well graded with a void content of 33% will use less lime than sand with a larger void content.
- Lime mortars are generally less wasteful than cement mortars because of their slower setting times and ability to be knocked up and used again.

ADVICE NOTE TO BUILDER

Q. What information about working with lime should be included in a brief with the tender documents for a new-build project? A natural hydraulic lime mortar is to be used for the building of both the outer brick leaf and inner concrete block leaf on a two-storey school.

A. Assuming it is the contractor's first experience of working with natural hydraulic lime and that all previous experience has been with cement, the contractor should be made aware of the following key points.

- Schedule work to allow for a more extended setting time than cement. The manufacturers of natural hydraulic lime give setting times for each of the four grades of NHL. Cold and/or wet weather extends normal setting times.

- Check restrictions on lift heights of newly built brick and block.

- Compared to cement, it will be necessary to protect the work from the elements for a longer period. Follow the manufacturer's instructions. After-care such as spraying with water and covering with damp hessian is often necessary to prevent the work from drying too quickly.

- Cement should never be added to the mix. Operatives who have worked previously with cement:lime:sand mixes need to be informed and monitored closely to make sure they stick to the specification.

- Mix ratios of natural hydraulic lime, sand and water need to be measured carefully to ensure consistency of mixes. Pre-gauged natural hydraulic lime and dry sand in silos is available and, where justified by scale, highly recommended.

- It is essential to allow for longer mixing times. Add water sparingly to keep the water content to the minimum. Additional mixing rather than additional water is the best.

- It is advisable to allow for some on-site training if the contractor's workforce has little or no knowledge of working with lime.

Silos for use with natural hydraulic lime, where large quantities are being used. Social housing scheme for Westmeath County Council, Kinnegad, County Westmeath

MATCHING NEW WITH OLD

Q. As an architect, I am designing an extension to an old stone building. The existing building has a lime wet dash which is flat in appearance and rather weathered. The new extension will be in concrete block. Is it possible to replicate the proposed new wet dash on the concrete block so that the new extension blends with the old?

A. Wet dash consists of lime and a coarse sand/aggregate. When new, the surface aggregate in the sand is hidden by a lime coating. Over time, the weather removes this lime coating and exposes the aggregate; it also flattens the overall surface. The colour, shape and size of the aggregate are clearly seen and should be matched in the new mix.

A natural hydraulic lime mix is used in three coats over the concrete block of the new extension. This is carried out as two laid-on coats (scratch and float) and finished with a thrown-on wet dash. The wet dash needs to be sufficiently wet (but not over-wet) to achieve a flat appearance. Too much water will result in cracking. Rather than trying to expose the aggregate on the new work, it would be better to apply a limewash to both old and new, creating an overall near-uniform look.

VOID TEST FOR SAND

Q. As a clerk of works, I need to know how to check the void content of sand on-site. How should I do this?

A. Most mortar mixes are designed for a 33% air content in sand. Two thousand years ago, the Roman architect Vitruvius referred to a 1:3 mortar mix — many modern mixes of cement:lime:sand such as 1:1:6 and 1:2:9 are based on one part binder to three parts sand. The binder content is designed to replace the air content. The binder, with consideration of the consequences (increased strength and loss of permeability), may be increased if the air content in poorly graded sand exceeds the norm. It is better not to have to do this and it may be worthwhile visiting the sand supplier's yard in advance of delivery to check that the air content of the sand being ordered is within the norm.

The following is a simple field test to check the air content of sand:

- A quantity of sand is first dried in an oven.
- The sand is placed in a container of known capacity.
- A measured quantity of water is added to the sand until it just begins to appear on top.
- The volume of the water added is divided by the volume of the sand and multiplied by 100. This gives the percentage of air in the sand.

The European standard for sand is EN 13139-2002.

WEAKER LIME MIXES – THEIR USE AND FUTURE

Q. Why use high calcium or the weaker pozzolanic or natural hydraulic lime mixes when they take longer to set and require more after-care?

A. The weaker lime mixes play a key role in the repair of old buildings when they are used to replicate existing mortars. High vapour permeability, flexibility, reversibility and the ability to fail before their host material are all key factors that support their use. In addition, weaker limes absorb more carbon dioxide than the stronger ones.

In a world dominated by Portland cement, concrete and mortar, there is concern that these weaker lime mortars have too low a compressive strength. However, this is not a serious concern when plastering over split-lath ceilings, mud-mortared stone walls, or low-fired clay brick. The main concern is their slowness to harden or set, but many jobs can be planned around this and very often the concern is not justified.

Workability is always important for the craftsperson. Sometimes it is so critical that the job cannot be done otherwise. These weaker lime mortars are generally the most workable of all the lime mortars.

When it comes to the repair of old buildings, lime mortars have relevance throughout the world. The author, Patrick McAfee (in white overalls), demonstrating the use of lime at a Stone Foundation symposium, Santa Fe, New Mexico

The control of water in the mix is important – 'less is best' is critical. The mortar mill (roller pan mixer) or the paddle mixer can produce highly workable mortar with a lower water content.

A slower setting time can be an advantage. It allows other work that cannot be rushed – such as decorative plasterwork – to be carried out.

There is little or no waste with slower setting lime mortars; generally, the mortars can be used a considerable time after mixing. High calcium limes, if kept under cover from access by air, can be used at any time in the future. *Knocking up* most of these mortars hours or days after first mixing produces an even more workable mortar.

NHL1 – a new addition to the existing range of natural hydraulic limes – arrived on the market in 2007 and should provide part of the answer to replicating the once common feebly hydraulic lime.

There are few craftspeople who are skilled and experienced in all areas of working with lime, but especially with these weaker lime mortars. As the strength of the mortar is reduced, increased skill is necessary. More

Part of a decorative garden path in
County Dublin. Coloured pebbles
were set into natural hydraulic lime
and sand mortar

Opposite Twelfth century chancel
with fifteenth-century stone vaulted
roof. A thick coating of cement
plaster was removed and the
ancient masonry repaired with high
calcium lime and brick pozzolan
addition and limewash finish

research is required in Ireland into our historic range of
weaker mortars, followed by training in their use.

Weaker limes have also found a bright new future in
combination with bio aggregates such as hemp. The high
vapour permeability of these limes makes it possible to
bind them successfully with hemp and other bio materials.
Not only are there benefits of low embodied energy, but
carbon dioxide absorbed by hemp is then sequestered
(locked in) by adding it to lime. The result is a carbon
neutral, sustainable, high insulating, environmentally
friendly building material.

GLOSSARY

aggregate: the hard filler materials, such as sand and stone, in mortars, plasters, renders, limecretes and concretes.

air lime: *see* high calcium lime.

allure stones: the cut stones laid to a fall, behind the parapet walls on tower houses. Provide a cover for the wall top, a means of taking water from the roof to spouts, and a pavement for walking on.

alum: aluminium potassium sulphate. Used as a fixative in colour washes applied to brickwork.

aluminates: compound of aluminium and oxygen. Becomes reactive when heated and gives a pozzolanic set when combined with calcium hydroxide.

arch barrel: the masonry vault of a bridge.

arris: a sharp edge at an external angle produced by the meeting of two surfaces.

ashlar: squared, regular masonry, with small joints.

autogenous healing: free lime (calcium hydroxide) in a mortar transported by water to fine cracks where it carbonates, thereby healing or closing the crack. Sometimes referred to as self healing.

ball of blue: added to limewash in the past to create an even more brilliant whiteness. A ball or bag of blue was associated with laundering clothes.

bastard tuck pointing: a form of tuck pointing applied to brickwork and commonly seen in Dublin, where it is also called wigging. The small, thin projecting tuck or ribbon is applied to or is formed from the uncoloured pointing insert or stopping. This surface of the stopping, each side of the tuck, is then coloured using a pigmented mortar to match the brick or the previously colour-washed brickwork.

batching: the process of adding the correct quantities of materials, either by volume or by weight, to the mixing process. When measuring by volume, a batching box or container should be used to ensure that the correct quantities and ratios of materials are achieved.

binder: the matrix between the aggregate particles in a mortar, render, plaster or limecrete mix. A paste of either high calcium lime, natural hydraulic lime or gypsum or a combination of these, with or without a pozzolan.

bond stones: generally used when through stones are not available. Laid in pairs, one opposite the other, overlapping, with their lengths running into the thickness of the wall.

brick jointer: a stiff, flat, steel tool with a wooden handle used in the tuck pointing of brickwork to apply the tuck or ribbon. Not to be confused with a flat jointer.

brick nogging: the brick and lime mortar infill to a timber stud partition. *See also* noggings.

calcining: the burning of calcium carbonate in a kiln to drive off carbon dioxide and produce calcium oxide.

calcium carbonate: limestone or shells composed of calcium oxide and carbon dioxide.

calcium hydroxide: calcium oxide (quicklime) combined with water.

calcium oxide: also called quicklime or lump lime. Calcium carbonate less its carbon dioxide content lost from the heat process of the kiln.

calp: a limestone particularly associated with Dublin. An argillaceous limestone (contains clay) extensively used for rubble walling and sometimes hidden behind façades of ashlar stone and brick. Was also burnt in kilns to produce lime, some of which was reported to be hydraulic.

carbonation: carbon dioxide combining with calcium hydroxide to form calcium carbonate.

chemical set: *see* pozzolan.

clamp fired bricks: bricks burnt in a kiln that has been constructed of the green bricks being burnt.

cloch aoill: translates from the Irish language as 'manure stone'. Manuring the land using quicklime or lime hydrate to increase alkalinity was the main purpose of limekiln construction and lime production in Ireland between the seventeenth and twentieth centuries.

cobbles: natural rounded stones, often glacial till, used as paving in farmyards, cattle byres, sheds and sometimes public areas for foot traffic. They were generally laid in earth. Cobbles associated with large eighteenth and nineteenth-century houses are sometimes seen laid in decorative patterns combining different coloured cobbles laid in lime mortar.

cold bridges: formed when there is poor thermal insulation between the external and internal faces of a wall.

colour wash: applied to brickwork only, and not to be confused with limewash. Composed of a colour pigment, water and a fixative, sometimes alum or even stale beer.

coping: the top stones laid on a wall. Often laid on edge to a variety of different styles. Prevent ingress of rainwater and consolidate wall tops, inhibiting plant growth and the loss of smaller stones from the top of the wall.

copperas limewash: *see* ferrous sulphate.

corbelling: courses of stone or brick, each over-sailing the one underneath so as to enclose a space. Strongly associated with Irish architecture over a prolonged period, from the Neolithic to the late medieval period.

core fill: stones laid in lime mortar in the centre of a wall. The term hearting is perhaps a more accurate description, as the centre of walls should not be filled but built carefully. *See* hearting.

crazing: small, fine shrinkage cracks in the finish or setting coat occurring when the plaster dries too fast. The previous coat (float coat) may not have been sufficiently damp. The scouring of the finish coat with a wooden float during its laying on prevents crazing. The finished work may need an occasional fine mist spray of water, and be protected from artificial heat, drying wind, sunshine etc.

culm crusher: a large edge-runner grinding stone drawn by a horse and used to temper culm (anthracite slack) and yellow clay (*dóib bhuí*). Used at times to temper quicklime and gravel. Found in the Castlecomer region only. *See also* roller pan mixer or mill.

darby: a long, flat board or straight edge with handles at the back at each end used by a plasterer to achieve flat plane surfaces on plasterwork.

daubing out: filling hollow areas on the face of a wall using a combination of lime mortar and stones, terracotta tiles or spalls of clay brick, prior to plastering or rendering. Also referred to as dubbing out.

devil float: a plasterer's wooden float with a number of small projecting nails. Used to scratch lightly the float or second coat.

disc cutter: a machine with a circular cutting disc powered by electricity or compressed air.

distemper paint: a paint generally used for internal work but sometimes used externally. It was made from whiting or chalk, size (animal glue) and pigment. Can be washable or non-washable. Colours are usually soft in appearance. The ingredients used to make distempers varied and could include a gum or casein (protein in milk) solution, oils, fats or resins, waxes and soft soap.

dóib (or *dóib bhuí*): An Irish term referring to yellow mud, a subsoil. Muds also could be blue/grey. *Dóib* was used for mortars, renders, plasters, mud wall building and puddle over arch barrels of bridges. Sand, gravel, lime, manure and other organic materials could be added at times.

double struck: a joint finish used mainly on brickwork. Carried out with the laying trowel or with a pointing trowel as the work proceeds. A slight projecting 'V' shape is formed flush to the face of the brickwork by striking each side of the joint.

dry stone: stone laid without mortar.

edge bedding: a sedimentary stone or certain metamorphic stones such as mica schist, slate, quartzite etc. laid with their natural bedding vertical. Seen on copings on wall tops. *See also* face bedding, natural bedding.

exothermic reaction: a chemical reaction that releases energy in the form of heat, as occurs when quicklime is added to water.

extrados: the outer curved face of an arch, arch barrel or vault.

face bedding: stones laid wrongly with their bedding planes vertical and running parallel to the wall face. These bedding planes or laminates can delaminate from weathering, and can burst from overhead compression forces. Because face bedded stones are not laid with their length running into the thickness of the wall, this type of work is structurally weak. *See also* edge bedding, natural bedding, face stones.

face stones: nearly all the stones seen on wall faces. These should be laid on their natural beds with their length running into the thickness of the wall. Vertical joints should be broken, one stone on two stones, and two stones on one stone.

fat lime: *see* high calcium lime.

feather edging: associated with pointing and plastering. The controlled working of a mortar away from its main bulk to a thin perimeter edge.

ferrous sulphate: sulphate of iron. A green crystal first mixed in warm water and then added to lime putty and cold water to make a yellow/ orange coloured limewash. Also referred to as a copperas limewash.

finish coat: the setting coat in plastering.

flat jointer: a flat, steel-bladed tool used in pointing stone and brick. Vary in width to suit different joint sizes. Used to lift, apply and compress the pointing mortar into the joints. Variously called flat bars, jointers etc. Not to be confused with a brick jointer.

float coat: the second coat in plastering.

flue liner: a hollow, fired clay unit used to line the inside of chimneys. Provides a uniform, smooth profile that is easy to clean and prevents the escape of fire and smoke. The flue liner replaced the previous brick built and parged (lime plastered) flue.

formulated lime: A hydraulic composed of natural hydraulic lime and/or calcium lime plus added hydraulic lime or pozzolanic materials including cement. The manufacturer must disclose its contents. To be included in the revised standard EN459 in 2010. Likely to be used in new-build. See also natural hydraulic lime and hydraulic lime.

Frenchman: a homemade tool used to cut the ribbon or tuck in tuck pointing of brick. It consists of a kitchen knife cut to a profile with a tiny, turned-down projecting blade at the end of the knife.

frogged brick: a brick with an indent or hollow in one or both beds. Associated mostly with machine-made bricks (after 1860s), although they were sometimes made by hand. The frog saved on brick clay, reduced the weight of the brick, and gave a key for the mortar.

galleting: the insertion of small stones, often of similar size, shape and colour, into the surface of soft mortar joints in stonework. Galleting looks and was intended to be decorative but it had a practical purpose: to stiffen up soft mortar, preventing mortar runs down the face of the wall. *See also* pinnings, sneck.

gauged brickwork: brickwork built with highly accurate, specially made, cut and rubbed bricks. Laid with very fine joints in lime putty and fine sand.

gauged stuff: generally referred to as plaster of Paris added to a lime putty and sand mix for plasterwork.

gauging: the process of adding a setting agent such as a pozzolan, plaster of Paris or natural hydraulic lime to a lime mortar to achieve a quicker set, hydraulicity or an increase in compressive strength.

'good hat and a good pair of boots': refers to a building needing a sound roof, rainwater goods and flashing, combined with a good base consisting of the efficient removal of rainwater well away from the building, good foundations and a ground floor higher than the external ground level.

harling: *see* wet dash.

hearting: the centre of a stone wall, between both faces, constructed carefully with stone and lime mortar. The stones are best laid transversely across the thickness of the wall. The first stones should be long and large, then medium-sized stones, followed by small.

high calcium lime: burnt from a near-pure limestone with no clay to produce calcium oxide which is then converted to calcium hydroxide (lime putty) or to calcium hydrate. Hardens through the process of carbonation. Also referred to as air lime. Calcium lime is not assessed on its compressive strength but rather on its lime content. CL90, the highest of these, is the one generally available in Ireland.

hot lime mortar: quicklime (calcium oxide) mixed with sand and water — a hazardous activity producing high temperatures and requiring special training and personal protective equipment. Used by stonemasons in the past as well as for backing coats in plasters and renders. Rarely if ever used today.

hydrate: just sufficient water added to quicklime (calcium oxide) to convert it to a hydrate or a powder. The hydrate can be high calcium or natural hydraulic lime.

hydraulic binder: a binder that can set under water. It sets and develops strength by chemical interaction with water.

hydraulicity: the ability to set under water, without carbon dioxide.

hydraulic lime: often used as a general term to include a number of different types of lime that have hydraulic properties. Specifically 'hydraulic lime' as designated in the proposed revision of EN459 c. 2010 is a lime with added materials such as cement, blast furnace slag, fly ash and limestone filler. For new-build. The manufacturer is not obliged to disclose its contents. See also natural hydraulic lime and formulated lime.

hygroscopic: a material that attracts moisture from its surroundings.

inclusions: pieces of unslaked quicklime seen in lime mortars, particularly hot lime mortars. These are undesirable, particularly in plasterwork, because of the danger of later expansion resulting in the popping of finished surfaces.

in situ: to make something in place such as construction carried out on a building site using raw materials.

intrados: the inner curve or underside of an arch, arch barrel or vault. *See also* soffit.

jacking: the lifting and pulling apart of masonry. Can occur from frost heave, hot lime mortars, oxidisation of iron, and plants such as ivy causing expansion within mortar joints.

jamb: the vertical surface of the side of a door or window opening

jointing: to shape, compress and finish the horizontal and vertical mortar joints of brickwork during the laying process. Usually carried out with a rounded steel bar. A modern method of joint finish mostly associated with cement mortars.

kiln-fired bricks: bricks fired in a permanent kiln as compared to a temporary clamp-fired kiln.

knocking up: the remixing and/or beating of mortars to increase plasticity.

laitance: the milk of lime or any other binder which, through working, is brought to the surface of a mortar joint, plaster or render. Steel tools will readily do this, sometimes to the detriment of the work. In such cases, steel tools are used sparingly and/or wooden tools such as the plasterer's float are used instead.

lead tray: a horizontal sheet of lead built into a chimney just above the roof. It projects and is turned up on the inside of the flue to catch rain which is then let out through weep holes.

lift: usually as in a lift of brickwork or stonework. When a lift is completed, scaffolding is erected so that the next lift can be built. A lift in stonemasonry is generally not more than 1.35 m. The slower set of lime mortars may restrict the vertical lift height that can be completed in any given time period.

lime kiln: a purpose-built structure to produce quicklime by the calcination of limestone. Traditionally, wood, turf or coal were used as fuel. The fuel was either mixed in alternate layers between the limestone, as in the running kiln (also called *tine seasta*, perpetual, continuous, commercial, sale or draw kiln), or kept separate as in the standing kiln (also called *tine aoil seasta*, French, flare or arch kiln).

lime putty: calcium hydroxide produced by slaking calcium oxide (quicklime) in water. Lime putty is sometimes produced by rehydrating high calcium bagged lime in water but this is not considered to produce as good a lime putty as the former method.

limewash: lime putty and water mixed together and applied with a brush. A pigment may be added.

mid-feather wall: *see* withe wall.

mortar: the mixture of a binder with sand and water.

mortar mill: *see* roller pan mixer or mill.

mud: *see* dóib.

natural bedding: the surface upon which the sedimentary stone was originally deposited. Applicable to sedimentary stones such as limestone and sandstone, but also metamorphic stones such as mica schist, slate and quartzite. Stones in a wall should be laid flat on their natural beds where they can resist both the forces of compression and weathering most effectively. *See also* face bedding, edge bedding.

natural hydraulic lime (NHL): a hydraulic lime produced by burning naturally occurring argillaceous and siliceous limestones. Available at present in three grades: 2, 3.5 and 5. These grades denote compressive strength in N/mm^2 at 28 days. (An NHL1 has recently arrived on the market.) NHLs are mixed with aggregates to produce mortars, renders, plasters and limecretes. For the repair of old buildings and new-build. See also formulated lime and hydraulic lime.

noggings: the materials used to infill and stiffen a timber stud partition to reduce noise and/or fire spread. Bricks laid in lime mortar were commonly used but also turf (peat). The partition would then be plastered with lime mortar on both sides.

non-hydraulic lime: *see* high calcium lime.

out of twist: a stone surface taken out of winding to a true flat plane by a stonecutter or stonemason.

parging: usually refers to the inside plastering of chimney flues using lime, sand and cow dung. Sometimes refers to the plastering of the underside of slates with lime, sand and animal hair where it is also called torching.

perforated brick: a brick with a series of round or square holes extending from one bed to the other. Associated with modern machine-made bricks. The holes provide a key for the mortar bed, reduce the weight of the brick, and save on clay.

permeability: how easily water can pass through a solid material.

perps: the perpendicular joints in a brick or stone wall.

pinnings: small stones inserted into mortar joints between larger stones to reduce the volume of mortar exposed to weathering. This is done after the larger stones are laid. Pinnings also stiffen up wall faces and prevent movement. *See also* sneck, galleting.

plug and feathers: a very old method of splitting stones still in use today. Two metal feathers are inserted into a drill hole, followed by a metal plug. A series of these is positioned in a straight line about 100 mm apart. The plugs are struck in turn until the stone splits.

Portland cement: limestone and clay burnt in a kiln to form a clinker and then ground finely with calcium sulphate added. Manufactured throughout the world. Used in concretes, mortars, plasters and renders for new-build.

pozzolan: reactive aluminates and silicates that combine with calcium hydroxide to produce a chemical or pozzolanic set in mortar. Similar to a hydraulic set.

pricking up: *see* float coat.

puddle: sometimes referred to as poddle. Mud applied over the arch barrel of a bridge to prevent lime mortar loss through leaching. Also applied to the walls or embankments of canals, and to contain water and prevent leaks.

putty: *see* high calcium lime.

quicklime: *see* calcium oxide.

rammer or ramming iron / bar: used for compressing deep mortar joints in masonry during the pointing process. They are homemade and of various styles. Some consist of a short section of flat steel bar welded onto a long round or square steel bar as a handle.

recessed joint: a mortar joint in brickwork or stonework purposely made by raking out, sometimes followed by jointing with a flat steel jointer. A type of modern finish that is not appropriate to older work.

Red Ochre: oxide of iron mined from natural deposits and ground, levigated (made into a smooth, fine powder or paste by grinding when moist) and dried before use. Roasting produces Venetian Red.

refractory cement: a cement that is resistant to high temperatures. When mixed with the appropriate sand, it makes a refractory mortar.

render: an external plaster applied in particular to stone walls.

reveal: *see* jamb.

ribbon: *see* tuck pointing.

roller pan mixer or mill: comprises two edge runner steel rollers/wheels contained within a cylindrical pan. The rollers or the pan revolve, resulting in lime and sand being worked together more effectively than just a simple rotational mixing process, as occurs in a cement mixer. High workability with reduced water content is the advantage. Also referred to as a mortar mill. *See also* culm crusher.

Roman cement: imported from Britain in the nineteenth century and used as a binder in mortar. Seen in repairs such as pointing to sea piers, lighthouses etc. but also in renders. Used also for casting decorative details on buildings. Very fast setting, it has a distinctive terracotta colour. Invented by James Parker in 1796.

roughcast: *see* wet dashing.

rubbing bricks: oversize bricks manufactured from a special clay to allow the brick to be cut and rubbed to accurate dimensions. These bricks were called rubbers and were laid with a fine lime putty mortar joint in gauged brickwork.

ruled or penny struck: a mortar joint with a rebated line jointed into the centre of flush or struck flush joints. Ruled or penny struck was executed either as a jointing or a pointing process.

sand: weathered, hard particles of rock, mostly silica, composed of particles less than 5 mm in size graded down to dust, ideally with no more than a 33% void content for mortars, renders and plasters.

scratch coat: the first or pricking up coat in plastering.

scudding: a thin, rich, very wet coat of binder and sand thrown onto a masonry surface to give a key for the first or scratch coat of a plaster or a render. Also referred to as spatterdash. Mostly used in modern work with cement mortars.

self healing: *see* autogenous healing.

setting coat: *see* finish coat.

setts: whinstone (basalt), rectangular in shape, cut and squared by hand and laid as road paving.

silicates: the salts of silicic acid; common in clay. Become reactive when heated and combined with calcium hydroxide.

slaking: quicklime added to water causes an exothermic reaction; the result is lime putty. Less water will produce a hydrate. This activity is hazardous and requires specialised training and personal protective equipment.

small tools: small plasterer's tools of various shapes and sizes. Used for intricate, delicate work, often with lime putty.

snapped headers: poor practice associated with eighteenth and nineteenth-century Flemish bond in brickwork where the length of the brick was purposely snapped in half and laid into the thickness of the wall rather than laying the full length of the brick. This saved on cost but sometimes resulted in the separation of the outer leaf of unbonded brickwork from the remainder of the wall.

sneck: a small stone that makes up the difference in height between two larger stones of unequal height, thus facilitating bonding. The sneck is structural, under compression, and is built in with the laying of the other stones in the wall. Snecks occur in precise formal patterns as in snecked work, or sometimes informally throughout rubble stonework. *See also* pinnings, galleting.

snots: a word in common usage within the wet trades to describe raggedy mortar projecting and overhanging past the face of the wall face, as in mortar joints.

soffit: the underside of an arch, arch barrel, vault or lintel. *See also* intrados.

souring out: the premixing of a high calcium lime, sand and water, and the storage of this for an extended period of time, possibly six months. The soured out mortar was kept undercover to prevent drying out and carbonation. The result, when used, was a highly workable mortar.

spalled: a brick or stone showing loss of material from its face, commonly caused by expansion from frost, oxidisation of iron or steel, salts, or plant growth such as ivy in mortar joints.

split lath: thin wooden laths split on the grain and fixed to wooden studs or ceiling joists with nails. They are fixed with gaps between the thickness of a forefinger to provide a key for the first coat of haired lime plaster.

stopping: associated with tuck pointing. Mortar, pigmented or unpigmented, inserted into a brick joint before applying the projecting tuck.

strap pointing: a projecting style of pointing wrongly applied to stone and sometimes brick, usually carried out in sand and cement mortar. It catches water, holds in damp and is very difficult to remove without damaging the arrises of the stone or brick.

struck joint: carried out during the process of laying brick and sometimes stone. The laying trowel is used to strike the mortar joint, creating a small ledge on the top horizontal arris of each lower brick or stone. Sometimes, these ledges were used to give a key for the first or render coat.

stuccodore: Italian for 'plasterer'. Commonly used expression in Ireland during the eighteenth century when plasterers from Italy and elsewhere were at work here.

stuccowork: high-quality external render such as ruled ashlar stucco.

swimming stones: stones that move after being laid. Often associated with wrongly laid, face-bedded stones, which can lead to collapse. Over-wet mortars and stones contribute.

through stone: a stone, the width of the wall in length, laid from one side of a wall to the other in order to provide structural strength to the wall. Laid at specific horizontal and vertical distances apart. In walls over 750 mm thick, these are unlikely to occur because of the difficulty in finding stones of such length.

torching: *see* parging.

trowelled down and trowelled up: occurs during the process of laying on the finish or setting coat in plasterwork. Laitance (rich lime content with little or no sand) comes to the surface and is then trowelled down with the plasterer's steel trowel. Finally, the work is trowelled up and finished, once again using the steel trowel.

tuck pointing: a thin projecting bead of white lime mortar, referred to as a tuck or ribbon, averaging about 3–6 mm wide and projecting 3 mm at most. Used to create the appearance of gauged brickwork, as well as to disguise a much wider mortar joint behind. Most tuck pointing in Ireland occurs in Dublin in a style known as bastard tuck.

Vegetable Black: a pigment used in colour washes to brickwork along with colours such as Red Ochre or Venetian Red. Produced by burning oils in a very restricted atmosphere. The clouds of carbon are passed through a flue and collected.

Venetian Red: *see* Red Ochre.

voussoirs: the stones or bricks forming an arch, arch barrel or vault.

weather struck and cut: a type of brick pointing popular in the twentieth century. On the horizontal joint, it is applied slightly weepered at the top arris and flush with the bottom arris.

weep holes: holes purposely left through vertical mortar joints in walls to release water from behind.

weeping stones: exposed, hard, dense stones that easily show condensation on their faces.

wet dash: an external render that is thrown on and left as-is. Called harling in Scotland and roughcast in England.

wet trades: refers to the trades of stonemasonry, bricklaying and plastering, all of which use mortars.

wiggers: bricklayers who specialised in pointing brick buildings in Dublin.

wigging: *see* bastard tuck pointing.

withe wall: the wall within a chimney that separates one flue from the other. It is important that they are bonded properly into the chimney. *See also* mid-feather wall.

Yellow Ochre: a natural mineral consisting of silica and clay that owes its colour to iron oxide.

BIBLIOGRAPHY

Books and Magazines

Allen, Geoffrey *et al.*, *Hydraulic Lime Mortar for Stone, Brick and Block Masonry.* Donhead Publishing, Dorset, 2003.

Allin, Steve, *Building with Hemp.* Seed Press, 2005.

Ashurst, John, *Mortars, Plasters and Renders in Conservation.* Ecclesiastical Architects' and Surveyors' Association, 2002.

Ashurst, John and Nicola, *Mortars, Plasters and Renders.* English Heritage Technical Handbook, Volume 3. Gower Technical Press, 1989.

Azbe, Victor, J., *Theory and Practice of Lime Manufacture: A Collection of Articles by Victor J. Azbe 1923-1946.*

Bevan, Rachel, and Woolley, Tom, *Hemp Lime Construction — a guide to building with hemp composites.* BRE Press, 2008.

Blundell, Cliff, *Precious Inheritance*: The Conservation of Welsh Vernacular Buildings, Quinto Press Ltd, 2007.

Bowen, Laura and Mathews, Nicki, *Reusing Farm Buildings, a Kildare Perspective.* Kildare County Council, 2007.

Building Limes Forum, *Lime News* Volumes 1-7 (1993-1999). Published privately.

Building Limes Forum, *The Journal of the Building Limes Forum Volumes 8 - 16* (2000-2009). Published privately.

(*The Journal of the Building Limes Forum, *initially called* Lime News, *is published annually. Some of the early volumes (2-4) included two separate publications within one year. The more recent volumes (8-16) are lavishly illustrated with photographs and drawings.*)

Collins, James, F., *Quickening the Earth, Soil Minding and Mending in Ireland.* University College, Dublin School of Biology and Environmental Science, 2008.

Conry, Michael, *Dancing the Culm, Burning culm as a domestic and industrial fuel in Ireland.* Chapelstown Press, Carlow, 2001.

Cowper, A.D., *Lime and Lime Mortars.* Reprinted by Donhead Publishing Ltd, Dorset, 1998.

Foresight Lime Research Team, University of Bristol, *Hydraulic Lime Mortar for Stone, Brick and Block Masonry.* Donhead Publishing, 2003.

Holmes, Stafford and Wingate, Michael, *Building with Lime: A Practical Introduction.* Intermediate Technology Publications, London, 1997.

Keohane, Frank, *Period Houses: A Conservation Guidance Manual.* Dublin Civic Trust, Ireland, 2001.

Leary, Elaine, *The Building Limestones of the British Isles.* Department of the Environment BRE, HMSO 1983.

Lynch, Gerard, *Brickwork History, Technology and Practice,* Volumes 1 and 2. Donhead Publishing, Dorset, June 1993.

Lynch, Gerard, *Gauged Brickwork, A Technical Handbook.* Gower Publishing, Aldershot, June 1990. Revised in 2006.

McAfee, Patrick, *Irish Stone Walls.* The O'Brien Press, Dublin. 1997.

McAfee; Patrick, *Stone Buildings.* The O'Brien Press, Dublin. 1998.

McDonnell, Joseph, *Irish Eighteenth-Century Stuccowork and its European Sources.* The National Gallery of Ireland, 1991.

Millar, William, *Plastering Plain and Decorative (1897).* Reprinted Donhead Publishing Ltd, Dorset, 2009.

Pasley, C.W., *Observations on Limes.* Reprinted by Donhead Publishing Ltd, Dorset, 1997.

Pavía, Sara and Bolton, Jason, *Stone, Brick & Mortar: Historical Use, Decay and Conservation of Building Materials in Ireland.* Wordwell Ltd, Bray, 2000.

Powys, A. R., *From the Ground: Up Collected Papers of AR Powys.* London, 1937.

Searle, Alfred. B, *Limestone & Its Products: Their Nature, Production, and Uses.* Ernest Benn Limited, London, 1935.

Semple, George, *A Treatise on Building in Water.* Dublin, 1776.

Vicat, L.J., *Mortars and Cements.* Reprinted by Donhead Publishing Ltd, Dorset, 1997.

Wilkinson, George, *Practical Geology and Ancient Architecture of Ireland.* Murray, London, 1845.

Wingate, Michael, *An Introduction to Building Limes. Information sheet 9.* Eyre & Spottiswoode Ltd, London and Margate.

Wingate, Michael, *Small-Scale Lime-Building. A Practical Introduction.* Intermediate Technology Publications, London, 1985.

Woolley, Tom, *Natural building, A Guide to Materials and Techniques.* Crowood Press Ltd, 2006.

Booklets and Articles

Byrne, Edward M., *A Guide to Lime and Its Uses In Conservation and Restoration of Buildings.* Published privately.

Byrne, Edward. M., *The Traditional Lime Company*. Published privately.

Conference Proceedings, COST and Historic Scotland. *Urban Heritage Building Maintenance, Lime Technology Workshop.* 1998.

Conry, Michael, *Culm Crushers, Edge Runner Grinding Stones for Tempering Culm.* Chapelstown Press, Carlow, 1999.

Department of the Environment, *Mortars, Pointing & Renders.* Conservation Guidelines No. 4.

Department of the Environment, *Decorative Plasterwork.* Conservation Guidelines No. 6.

Department of the Environment, *Brickwork & Stonework.* Conservation Guidelines No. 8.

Georgian Group Guides, *No. 5 – Render, Stucco and Plaster, A Brief Guide to the History and Maintenance of Georgian Renders and Plasters.* The Georgian Group.

Grandison, L.S., *The Nearly Non Technical Book on Plasterwork.* Peebles, Great Britain, 1998.

Historic Scotland, *International Lime Conference, Lime News, Vol. 4.* The Building Limes Forum and Historic Scotland, 1995.

Historic Scotland, *Earth Structures and Construction in Scotland:* Technical Advice Note 6, 2002.

HL2. *Blue Lias Hydraulic Lime, Technical Background and Best Practice Instructions for General Masonry and Render.* Published privately.

Holmes, Stafford, *Evaluation of Limestone and Building Limes in Scotland:* Research Report, Historic Scotland, 2003.

Induni, Bruce and Liz, *Using Lime.* Published privately.

Leslie and Gibbons, *Scottish Aggregates for Building Conservation.* Historic Scotland, 1999.

Lochplace. *Building Conservation.* Published privately.

Lynch, Gerard, Roundtree, Susan, and Shaffrey Associates Architects, *Bricks, A guide to the Repair of Historic Brickwork.* Department of the Environment, Heritage and Local Government, Government Stationery Office, 2009.

Lynch, Sean, *The Stuccowork of Pat McAuliffe of Listowel.* Pamphlet, Siamsa Tire, Tralee, County Kerry, 2009.

Mack, Robert C. FAIA, and Speweik, John, *History of Masonry Mortar in America 1720 – 1995.* Technical Preservation Services, National Parks Service, U.S. Department of the Interior.

Mack, Robert C. FAIA, and Speweik, John, *Repointing Mortar Joints in Historic Masonry Buildings.* Technical Preservation Services, National Parks Service, U.S. Department of the Interior.

Meek, T., *How to Harl a Building with Lime.* Published privately.

Miniere Di San Ronedio, *Natural Hydraulic Lime.* HD System.

Narrow Water Lime Service, *Harnessing Traditional Materials & Techniques for Conservation and Restoration.* Published privately.

O'Keefe, P. and Simmington, T., *Irish Stone Bridges: History and Heritage,* Irish Academic Press, Dublin, 1991.

Rourke, Grellan D., *A Study of Historic Lime Mortars at Ardfert Cathedral, Co. Kerry, Ireland: the re-creation and use of replica lime mortars for its conservation.* The Journal of the Building Limes Forum, Vol 14, 2007.

Schofield, J., *Basic Limewash.* SPAB Information Sheet No. 1. Reprinted 1991.

Schofield, J., *Lime in Building — A Practical Guide.* Revised edition. Black Dog Press, Great Britain, 1999.

Scottish Lime Centre, *Case Study of Traditional Lime Harling.* Discussion Document 1996.

Scottish Lime Centre, *External Lime Coatings on Traditional Buildings.* Technical Advice Note, Historic Scotland, 2001.

Scottish Lime Centre, *Preparation and Use of Lime Mortars.* Revised edition. Historic Scotland, Edinburgh, 2003.

Scottish Lime Centre, *Preparation and Use of Lime Mortars:* Technical Advice Note, Historic Scotland. 2003.

Simpson and Brown, *Conservation of Plasterwork.* Technical Advice Note, Scottish Lime Centre, and Historic Scotland, 2002.

Society for the Protection of Ancient Buildings (SPAB). *Repointing Stone and Brick Walling.* Technical Pamphlet 5.

St Astier, *Pure & Natural Hydraulic Limes.* Published privately.

Williams, G.B.A., *Pointing Stone and Brick Walling.* SPAB. Reprinted 1986.

Williams, Richard, *Limekilns and Limeburning.* Shire Publications Ltd, UK, 1989.

INDEX